Donna Sessions

ALTERNATIVE AND COMPLEMENTARY THERAPIES

by

M. Mirchandani, MD

Published in USA by
Traderoute Publications

ALTERNATIVE AND COMPLEMENTARY THERAPIES

Published in USA by Traderoute Publications under agreement with M. Mirchandani

DISCLAIMER

The author, publisher, endorsers, distributors, companies, or anyone else whose name appears in this book shall have no liability or responsibility to anyone with respect to any loss or damage caused, or alleged to be caused, directly or indirectly by the information contained in this book.

This information is not intended for the diagnosis, treatment, cure, or mitigation of any illness, disease, or medical condition. The content of this book is for information only, and its application is not intended to encourage self-treatment by the layperson. Any application of the information and/or recommendations in this book is at the reader's own discretion and sole risk. Please see a qualified health care professional for medical advice before taking any therapy or any other nutritional supplement.

Editors ~
Floyd Hillman
Barbara Hutchings

Cover Design ~
David Tunes

Indexer ~
Barbara Tunes

First edition: *June 2002*
Second edition: *May 2004*

USA Publisher:
Traderoute Publications
517 E. Broadway, Suite #9
Helena, MT 59601 USA
www.traderoute.info

ISBN: 0-9719916-0-X

FORWARD

After 30 years of successful medical practice, during the mid-nineties, I realized that traditional treatments for the degenerative diseases of ageing did not slow the progression of the underlying disease process. It was frustrating to see patients stricken with various degenerative diseases not evincing any improvement despite protracted treatment, particularly those with cardiovascular diseases resulting from reduced blood flow to vital organs due to atherosclerotic plaque within the arteries. Many patients of heart attack, stroke, peripheral vascular disease, angina, senility, arthritis, diabetes, cancer, Alzheimer's and Parkinson's diseases, etc., continued to be in distress even after surgery and other traditional treatments. These situations, besides leaving me extremely helpless, seemed almost to contradict my calling as a doctor which dictated that I employ every faculty at my disposal to assuage suffering and bring some sort of relief to my patients.

When Dr. Moti Mirchandani first introduced me to Chelation Therapy, I discovered that, far from being skeptical, I was really excited by this novel and apparently very successful approach to health care. After an extensive and in-depth study of the scientific literature I was firmly convinced and introduced these principles to the treatment protocols of my many patients. That all of them showed improvements, ***from the remarkable to the astounding***, surprised even me! Subsequently, I diversified into other alternative therapies like Plaquex, Dimethyl Sulfoxide (DMSO), Hydrogen Peroxide and Ultraviolet Blood Irradiation (UVBI) and successfully treated various diseases that had earlier proved to be stumbling blocks. My experiences with the new medicines and therapies are similar to what Dr. Mirchandani saw and experienced first-hand in Germany and Switzerland and which are described in this book. My emphasis today is on early detection and prevention of disease, including life style changes, proper nutrition and nutritional supplements. Some of these therapies are also being given as a preventive measure.

I routinely see reversal of cardiovascular diseases using a comprehensive programme of Chelation, DMSO and other therapies which usually eliminate any indication of vascular surgery and results in less reliance on prescribed medicines. I have also seen improvements in other diseases of ageing such as arthritis, diabetes, etc., all of which are simply elucidated in this book. Read about and understand your body so as to take the right decisions about enhancing health and improving the quality of your life. You may be able to avoid bypass surgery, slow the ageing process and maybe get cured of the so called incurable diseases. I recommend this book as an educative guide to all, regardless of creed or status, so that your future health and ability to ward off disease are enhanced.

Dr. O. P. Agarwal

M.B.B.S., MD., Ph.D., D. Sc.
Senior Consultant, Non-Invasive Cardiologist
46, Jawahar Park Market, Bareilly-243001, India

Introduction

I am sure that my story is not very different from the millions of those who have also had the misfortune of being struck down by a heart ailment. In September 1992, while relaxing at home after a hard day's work, I was struck by a heart attack without any warning. One moment I was sprawled comfortably on a sofa, and the next I was hit by this sledgehammer blow in my chest which left me writhing in pain and gasping for breath. I was rushed to the hospital where a battery of tests confirmed that not only did I have triple vessel disease, but had suffered a major myocardial infarction as well. The local specialists insisted that I undergo an immediate bypass surgery which, they iterated, was the only way of relieving the undue strain on my heart. I'm not too sure what made me refuse the operation then, and I can only put it down to a fear of the unknown in general and to surgical procedures in particular. Anyway, against the wishes of the medical authorities ***and*** my family, I got myself discharged and returned home. I was prescribed some essential medications, which I took religiously, even though I did not understand their functions. In addition, on doctor's orders, my wife put me on a strict diet and I totally gave up smoking. To say that this turned my regular routine topsy-turvy would be making an understatement.

At that time, I was stationed in Assam and enjoyed a fairly successful career. I was the engineering director of five companies whose various occupations ran from civil engineering and construction works to fabrication and the tea trade. Though I was extremely happy with the working conditions and the rewards thereof, I must admit that mine was quite a high pressure occupation that kept me busy with a 12-hour working day—sometimes seven days a week! Of course, that hectic work routine came to an abrupt halt after my heart attack. About a month later, frustrated with the restrictions on my normal activities, I went off to the USA in the hope of finding some solution to my problems. Once there, I underwent a series of clinical tests that did not corroborate the back-home hospital findings. The American cardiologist advised me to continue with medication only, adding that bypass surgery was not indicated at the moment. One day, while on a flight from Atlanta to Los Angeles, I got to talking to my American seat mate and discovered that he had gone through the very problems that were plaguing me. He also made the startling disclosure that he currently subsisted purely on vitamins and herbs and had stopped all prescribed allopathic medication. I was amazed at what he had to say and literally pestered him with questions throughout the flight, finally eliciting the information that I should get hold of a book entitled *Nutritional Healing*.

Immediately upon landing at LA, I was about to make a beeline for the nearest bookstore when my new-found friend tapped me on the shoulder and presented me with his copy. He disappeared into the crowd before I could even thank him properly. I avidly read through what the book had to say and was astounded at the detailed beneficial effects of various vitamins, minerals and herbs. Based on the information and instructions provided, I started myself on these nutrients and soon found my health improving in leaps and bounds. After three months of treating myself thus, I reached a stage of physical well-being where it seemed that there was absolutely nothing wrong with me.

In the meanwhile, spurred on by what I had read in *Nutritional Healing*, I realized that information regarding the human body and the influx of diseases had interested me very much, and also that a smattering of medical knowledge was essential if a layman was to understand how and why things went wrong. So I started studying some basic medical literature, concentrating, at that time, mostly on information regarding the heart. Of course, one thing led to another and soon I found myself almost totally immersed in advanced medical studies, and the more I read the more I wanted to know. I educated myself to the extent that I was able to make pertinent conversation with and even question qualified medical personnel. My engineering background certainly helped tremendously in making me understand not only the working of the human body but also the 'how and why' of the ill effects of various diseases. By the time I returned to India, I had regained much of my health and medical tests showed that I was in good shape. However, I was advised to avoid stressful scenarios which could prove harmful in the future. As a result, I resigned from my posts in Assam and shifted to Delhi where I spent the next two years constructing my house. This break also allowed me to enjoy time with my family and participate in the growing up of my six year old son—an enjoyment that had not been possible with my earlier lifestyle. It goes without saying that I continued with my medical self-education.

In 1994, I had gone to Germany to sign an agreement for setting up a plant to manufacture orthodontic equipment. I must add here that I have a host of relatives in Germany all of whom belong to the medical fraternity. It also goes without saying that, since the condition of my arteries had neither improved nor deteriorated further, all my relatives kept insisting that I undergo a bypass surgery which would be carried out under their personal supervision. Though I repeatedly declined the 'invitation', the pressure never let up. One evening at a house party, which was more or less a gathering of doctors, the topic of heart disease cropped up, with me being pointed out as a prime candidate for bypass surgery. Since all ***my*** family were of one mind in this context, I walked over to a renowned German surgeon and asked, "Why hasn't any oral medication or other means been invented whereby arterial deposits can be dissolved and removed from the body in a natural manner?" Imagine my incredulity when he calmly replied, "Oh yes, such a remedy **is** available. In fact, I even happen to know a doctor who regularly carries out these treatments". I lost no time in finding out who and where this doctor was and reached his clinic almost in a flash. What surprised me was the sight of about twenty patients undergoing IV drip simultaneously—all of them looking very calm and reassured. The doctor was extremely courteous and explained, as best he could in broken English, what the procedure was and how the medicine worked. ***Every single one of the patients there had arterial blockages of varying degrees and all of them were positively improving with each succeeding treatment.*** This was my first introduction to Chelation Therapy and I just couldn't believe what I was witnessing or what I was hearing. The German therapist very kindly gave me some literature to read as well as addresses of various chelation doctors in the USA.

Within a few days of my talks with the German chelation doctor and after having read the book *Bypassing Bypass* by Dr. Elmer Cranton, MD, I took a flight back to the States where I met with over a dozen qualified practitioners of this new therapy in various cities and gained a more intimate insight into this absolutely marvelous treatment. Having collected as much information as possible, I eventually arrived at my brother's place in Honolulu for an extended stay. It was here that I was introduced to Dr. Ronald Sorenson who explained further the intricacies of chelation to me. Once I was fully convinced about the veracity of the treatment and all my queries had been satisfactorily answered, I had my very first chelation session. Dr. Sorenson was quite intrigued with the extent of my medical knowledge and, both during and after the twenty treatment sittings that I had with him, decided to teach me how to carry out chelation therapy. I purchased and read many books on various alternative therapies and was given access to hundreds of published studies by Dr. Sorenson. It was then that I started building up my own library of recorded facts and patient histories.

Towards the end of 1994 I returned home, having carried with me sufficient medicine for ten more chelation sessions—which is where I became totally frustrated! I must have approached at least fifty doctors in Delhi for the required treatment. None of these doctors had either heard of chelation or even knew that such a therapy existed! I was getting fairly desperate when I luckily met Dr. Ramesh Bhola. Not only had Dr. Bhola heard of chelation, he was also quite eager to acquire any of the latest information. He agreed to carry out my treatments but confessed that he had no knowledge about the protocol. Under my guidance, Dr. Bhola was able to complete my remaining treatments. I had, by now, discarded the use of all allopathic medication and continued only with vitamins, minerals, herbs and a controlled diet. Having carried out some necessary lifestyle changes, I assure you that health-wise I behaved and felt like an energetic twenty-year old! It was about this time that I decided to promote these therapies in India as a viable and successful alternate to bypass surgery and so, with Dr. Bhola as an active participant, my Alternative Medical Centre came into existence. Having successfully chelated a few patients, I now made regular visits to Europe and the USA to carry out further research, attend alternative therapy seminars and also gather the latest information.

Between 1994 and 1998, in Europe and India, I had personally carried out thousands of treatments and, besides chelation, also gained experience in the other forms of alternative therapies such as plaquex, hydrogen peroxide, ultraviolet blood irradiation, hyperbaric oxygen and dimethyl sulfoxide. As and when a particular therapy was perfected abroad, I lost no time in introducing the same back home in India. By now, I had also established myself as a recognised expert on alternative medicine and continued to enjoy an excellent rapport with some renowned practitioners both in Germany and Switzerland. During one German seminar in 1998, I was granted an Honorary MD degree in Alternative Medicine and, as a result, was permitted to enroll in the Alternative Medical Course of the Indian Board of Alternative Medicines. I successfully completed the curriculum and obtained a Doctor of Medicine Degree, which is affiliated with The Open International University for Complimentary Medicine.

Despite the current proliferation of alternative medicines and therapies in Europe, the USA and many other advanced nations, the same are still relatively unknown in India. To spread the 'good news', I decided to produce this informative guide and hope that through these pages the uninformed layman will gain valuable insight to the miracle therapies that are literally passing him by! Since I am living proof—we have hundreds of other such successful records—of the wonders that these marvelous medicines have wrought, I do not hesitate to highly recommend these perfectly safe and truly effective treatments to one and all. Chelation has cleared up most of my arterial blockages; plaquex therapy has enhanced my kidney function and made me feel younger; DMSO has healed wounds, cleared up skin rash and made pain vanish;

hydrogen peroxide has strengthened my gums and teeth; ultraviolet blood irradiation has removed toxins from my body and cleared up an irritating cough. These are but a few of the beneficial changes that I have experienced personally, but I can also give you numerous examples from the hundreds of patients whom I have successfully treated. More than just a guide to the various therapies, this book will also enlighten you on the working of your body and goes on to explain how we inadvertently tend to mistreat ourselves through ignorance. As has often been said, today's deadlier maladies are all "diseases of modern living". There is a need for all of us to afford some sort of protection ***to ourselves***. This can only be achieved through knowledge, which is the essential means for the promotion of good health and healthy living.

Information and understanding about health and medicine is expanding exponentially every year. The best I can do is put forth information as it is currently known. This does not guarantee that the information in this book is completely accurate or safe. This book is therefore an educational guide only, and should not be used as a substitute for the advice and care of your personal physician.

Although it is my intention to educate health conscious consumers and motivate them to take a more active role in preventive measures that promote vitality and longevity, no book, including this book, can replace the services and knowledge of a trained health care professional.

I make no claims of originality. The knowledge gathered from numerous sources along with experience gained through many years of practicing Alternative Medical Therapies by healing numerous patients, has been collated to form this publication.

All views expressed are those of the author and may differ from the Medical Community.

Moti Mirchandani, MD/AM, Ph D
Fellow-WSAM; Member-IIHT
Alternative Medical Center,
C-6/2, DLF City, Phase I,
Gurgaon – 122 002
Haryana, India
E-mail: mirc@vsnl.com

Contents

CHAPTER 11

CHAPTER 12

I AM FINE - THANK YOU !

There is really nothing the matter with me.
And I'm actually as healthy as I can be.
I have arthritis in both my knees,
And when I talk, I talk with a wheeze !
My pulse is weak and my blood is thin
But, I'm awfully well for the shape I'm in !

My real problem is that my "Get up and go",
Got up and *went* - to my utter chagrin and woe.
But before going he winked and whispered in my ear
And I had to strain hard because I really can't hear.
"Beware", he said, " because folks may laugh at you,
Don't even mention some of the places we've been to."

Arch supports I have for my feet
Or I wouldn't be able to walk in the street.
Sleep is denied me night after night,
But every morning I find I'm all right.
My memory is failing, my head's in a spin,
But I'm awfully well for the shape I'm in !

Old age is golden - I have heard it said,
But I often wonder as I get into bed.
With my ears in a drawer, my teeth in a cup,
And my eyes on the table until I wake up !
Before sleep enfolds me I always whisper to myself,
Is there anything else I should have laid on the shelf ?

I come down in the morning and dust off my wits,
Pick up the paper and read the 'Obits'.
If my name is missing, I know I'm not dead
So I have a good breakfast and get back into bed.

The moral is, as these truths I unfold,
That for you and me who are growing old,
Its better to say, "I'm fine" with a grin
Than to let folks know the true shape that we're in !

CHAPTER 1

TRUE HEALING - THE CURE FOR ALL MALADIES

Oh, Woe Is Me!

Mr. JS was truly at his wits' end. At 64, the ageing process was a heavy burden, and the years sat upon him like the proverbial Sword of Damocles. Eight years earlier he had contracted coronary artery disease which resulted in clogged arteries and eventually a heart attack. His normal lifestyle became severely curtailed thereafter. In order to remain physically functional, and as advised by the doctors, he underwent bypass surgery. This in turn necessitated a reliance on drugs and other heavy medication to stabilize his potentially dangerous health condition—a process that kept him functional but physically weak. Unfortunately, contrary to the specialist's assurances, the bypass ***did not*** solve the 'heart problem'. Instead of attaining even a minimum level of post-operative fitness, his doctors' visits became even more frequent and, on many occasions, he was either subjected to re-hospitalization or endured repeated, extensive and exhausting clinical investigations.

Unlike the plague or other infectious diseases, a heart ailment is a very serious problem and will not simply 'just go away'. An individual should be wary about being persuaded into believing that surgery, whatever the procedure, is the *be all and end all remedy*, regardless of the eminence or experience of the propagating source. Seven years after the bypass, JS' arteries again became clogged and this time around, to un-block and facilitate normal blood flow, he was subjected not only to a **balloon angioplasty** but also had to be fitted with a **stent**. Both of these procedures are invasive, and it was akin to going through surgery again. Convalescence entailed further heavy medication—mostly drugs which played havoc with his already weakened condition instead of providing recuperative benefits.

The truly crushing blow came six months later when, despite being meticulous with his medication regimen, the arteries started calcifying anew giving rise to fresh blockages. Since there seemed no other recourse at the time, he suffered through another invasive procedure known as **angiogenesis** which entails gene therapy. ***What's next?*** You guessed it, a whole host of additional medication and drugs! The subsequent nine months of a drug-dependent existence had literally reduced JS to a nervous wreck. He suffered from hypertension, was almost completely bed-ridden, couldn't move or walk without pain and even the slightest exertion caused severe chest pains and breathlessness. He required heavy medication for almost every breath that he took; in addition to which he was consuming *10 sorbitrate tablets a day!* But worse was yet to come. His doctor finally gave up and conceded that nothing more could be done. Was ***THIS*** how he would spend the rest of his life!

After the heavy expenses, the physical trauma, the pain and the distress that he had endured, would dark clouds continue to loom ominously over JS' life? Was he truly doomed to spend the rest of his life in this semi-crippled manner? Was this, in fact, 'The End'? Certainly not! A silver lining appeared with his, purely by chance, lucky discovery that there existed other ***effective*** and ***life-saving*** alternative therapies which were also readily available. From that very first session of new therapy at our clinic, JS' sun shone magically again! More heartening, perhaps, was the assurance of no more surgery or hospitalization. Invasive procedures were neither part nor parcel of ***this*** therapy and the painless treatment would be accomplished in relative comfort in an out-patient setting. Heavy medication was largely done away with. The therapy was **much** cheaper and—*more importantly*—to his utter delight, the treatment assured positive benefits as opposed to merely being a short cut to relief. Suffice it to say that nine '*miracle-therapy months*' later, JS was on his feet again and resumed all normal activities without any undue stress or strain. Due to his age, and the fact of previous surgical procedures, he perforce *has* to continue with certain essential heart medication and must consciously take precautions to avoid overstraining himself. All this notwithstanding, the feistiness is definitely back in his step as is the smile on his face!

Miracles? In Reality?

Definitely so! However, before we proceed with our educative foray, there is the need for simple introductions to familiarize the reader with the therapies which will be discussed further in this book. Reality assumes a palpable form through easy descriptions, thus making the understanding and awareness of the basic functions that much easier.

These truly miraculous new age therapies are medically, scientifically and more popularly recognizable as:

Chelation Therapy. (Pronounced key-lay-shun) This is a safe and effective treatment for restoring blood flow in patients suffering from hardened and blocked arteries. It removes the undesired accumulation of metallic elements in the body while normalizing their correct distribution throughout the body. Chelation is a healing agent for cancer, diabetes, gangrene, high blood pressure, arthritis and other diseases of ageing. It entails the intravenous infusion of a synthetic amino acid called EDTA (ethylene-diamine-tetra-acetic acid).

DMSO Therapy. Dimethyl Sulfoxide (DMSO) is a simple by-product of modern paper making which, in liquid form, has a garlic-like odour and resembles mineral oil. It can be administered orally, intra-muscularly, intravenously or simply applied to surface skin. DMSO is one of the finest topically applied painkillers and is a potent anti-inflammatory plus an anti-arthritic remedy. It has forty major therapeutic properties, which act beneficially in the process of healing traumas/injuries to the body and the brain and also helps to cure and reduce the ill effects of various diseases.

Lipid Replacement Therapy. Also known as Plaquex Therapy, this is a developed generic product isolated from soy beans which reverses age-related damage to the body by revitalising cell membranes and helping the essential cells of the heart, lungs, kidneys, eyes, skin, liver and most other tissues and organs of the body to 'rejuvenate'. It is infused intravenously.

Ultraviolet Blood Irradiation Therapy. This therapy is administered by drawing 50 - 60 cc's, or more, of venous blood through a special tube placed in an ultraviolet light generator and then returning this same blood back to the body. In this closed system, the irradiated blood stimulates other blood cells as it moves through the circulatory system. This action results in an increased number of both red and white blood cells, increased hemoglobin, increased lymphocytes, decreased blood viscosity and many other benefits.

Hydrogen Peroxide Therapy. This is the method by which oxygen is caused to be dissolved in a serum. The peroxide then acts as a transporter carrying this oxygen throughout the body, where the oxygen-serum combination is then released at various points. The peroxide maintains membrane integrity and also participates in the transport of some substances across membranes—acting like insulin by affecting the RBC (red blood cell) membrane to transport glucose.

Alternative Medicines. Literally, alternative medicine means alternate to something else, the 'something else' in this case being allopathic—or western—medicine or orthodox medicine! In 1973, at the very first convention of the World Congress of Alternative Medicine held in Rome, no less than 135 different therapies were programmed for discussion. The experts of the World Health Organisation have identified and enlisted more than 100 types of practices.

HALT! Who Goes There?

Having reached this far, with tongue-twisting names and descriptions tending to indicate bodily functions, treatments and medication, we are sure that you have been forced into assuming "definitely a medical book. Not another dreary one?" Right? WRONG—*and on all counts!* This book is *neither* conceived *nor* is it written as any sort of complicated medical textbook. Special efforts have been made to keep it simple, understandable and easy to read. It is meant as a means of both garnering and passing on information that is either not readily available or which has been so couched in medical terminology as to make no sense at all to the layman. The ultimate goal is to enable every individual, regardless of qualifications, degrees *or* the extent of medical know-how, to have through enjoyable reading, the chance of assimilating valuable insight into some of the latest therapies and medications which are fairly new in India. A comprehensive, all-informative compendium on these treatments has not yet been published. Therefore, through these, the pages of our first-time effort, we look forward to the speedy spread of the 'good news'. During this educative process, it will be necessary to fall back on some detailed medical descriptions and procedures, as well as a bit of history, for you to better understand and appreciate what is being conveyed. However, these will be totally lucid and have been made as concise and uncomplicated as possible.

While preparing this manuscript, we have referred to numerous patient reports from various doctors world-wide, all of whom have practiced, and continue to practice, alternative therapies. Like JS, there were literally thousands of others world-wide who were also driven to despair, not only by the various and sometimes baffling maladies that afflicted them, but by the utter and complete disruption caused to their personal lives as well. The frustrations of seemingly never being able to get well, the treatments undergone notwithstanding, were almost epidemic in proportions. And like JS, *all* of these patients regained varying but *very positive* degrees of 'functional control' over their health, thanks—in no mean way—to the various 'miracle' alternative therapies.

Every one of these therapies has been practiced abroad at some stage or other, albeit heavily doused in many a controversy, for over a century! You may be somewhat nonplussed to know that they are actually *old* in origin but are being termed *new*. The reason is that, though they may have been in existence for a long time, their true efficacy and reliability in the treatment of really chronic cases has only come to light during the last 40 years or so. Then again, they have had to face stiff opposition from most of the already established and accepted norms of medicine. This has been the main hurdle to the progress of these therapies. Since they are fairly recent in India, it is hoped that this book will create enough curiosity to ensure the successful spread of information. There is a genuine need for a wider recognition and appreciation of these curative practices and beneficial health therapies which, besides being highly scientific, have also proved their efficiency in overcoming most—if not all—of the major medical dysfunctions that plague modern society.

The need of the hour is a viable form of self-education in respect of those deadlier 'enemies' of our health and well being, nay, of our very lives! Most of these foes do not manifest themselves overtly but lie in wily ambush, biding their time and waiting to pounce when least expected. These pages are intended to 'light the way' for an easier recognition of these 'villains', to understand better the *How's* and *Whys* of the havoc being perpetrated on human existence and to elucidate the methods of effectively enhancing our defensive mechanisms. If you happen to be young, *please do NOT*, not even for a moment, make the mistake of thinking that only the older, and probably physically weaker, section of humanity is the more likely prey. Exhaustive studies have confirmed that these 'enemies' do not recognise age and if offered a target, relentlessly press home their attacks on anyone and everyone. The brighter side to this grim picture is that it *is* possible to accord effective 'protection'. However, the best remedy of all may be summed up as "***Prevention*** *may certainly be the better part of* **cure,** *but for prevention to be really effective one needs necessarily to* ***know, recognise and then***

take appropriate action". One simply *cannot* go through life depending solely upon miracles!

So, What's the Hue and Cry About?

Are you finally figuring things out? Wonderful! Please keep your wits about you because we're going to fire up some more enthusiasm. If you somehow develop a grouse, hang on there for a minute; we've written plenty for you to let off steam about as this progresses! For the present, we need a teensy bit of your undivided attention. As Caesar might have put it, "....countrymen, lend us your ears"—and *eyes* we add!

Now don't get hyper; not only does your body have many hues, it also has plenty to cry about! So, where to next? Let's try the word '*disease*' for openers. Know anything about it, besides its also being another term for '*bad habit*'? Probably a better description would be 'disability'. If it doesn't work, it is ***disabled,*** it is ***diseased***! So too with our anatomies. To truly understand the ill effects of disease, we've got to return to that smallest factor of living tissue—*the cell.* There are over 100 trillion cells in the average human body and the health of the 'whole' depends entirely upon just how well each one of these individual cells is working.

Each cell is actually a living chemical factory, a miraculous and precisely accurate production unit. The cell extracts nutrients from the food we eat and oxygen from our lungs to be able to produce energy—the energy for life. Each cell is interconnected and at some, yet to be understood, unconscious level, the brain is certainly aware of what is going on in all of them. Though knowledge of their existence and function has been available for quite some time, it is only recently that an understanding has emerged of what goes wrong when a cell goes out of control, or malfunctions, or clogs up. This phenomenon of a cell going haywire—'out of whack' or 'out of sync' in modern parlance—is the primary reason for the onset of any disease. For example, when a cell goes out of control and starts wrongly multiplying and proliferating—you have cancer! *We're deadly serious about this.* It needs a very vivid imagination to mentally visualize the actual size of one of these less-than-microscopic guys who freely inhabit our insides, without whom we could not exist and any single one of whom—in a maniacal twitch of inexplicable behaviour—is capable of sending each one of us literally into orbit, sans the help of rockets! We sincerely hope that you get the message: These chaps have to be handled with kid gloves! Despite the ominous overtones, there's really no call for alarm – YET! So long as this has made sense up until now and your curiosity has been piqued enough, all you have to do is treat *this* alarm as a 'wake up' call in the correct perspective. Knowledge has no limitations; it is the imparting that's difficult.

It is important to remember that cells vary greatly in structure and function. The cells of the skin, the linings of the intestinal tract and the blood are continuously being worn out and regularly renew and replace themselves through division. Other tissues, such as the brain, nerves and muscles—*including the heart*—are made up of non-dividing cells that, once worn out, cannot renew themselves. Over time, these non-renewable cells become increasingly damaged in the course of their activities; they age and eventually die clogging tissues, organs and biochemical pathways as cellular 'rubbish'. When your cells are damaged, ***you*** are damaged. When your cells perform inefficiently, ***you*** perform inefficiently. And when a sufficient number of your cells *die— ?* We leave the rest to your imagination!

CHAPTER 2

THE FUNCTIONS OF WHAT'S INSIDE

You Belong to My Heart – Now and Forever!

At this juncture we are going to say 'hello' to and get intimate with one of the most important organs in our body—the heart. There are some who might argue for the brain's greater importance, but we beg to differ. The simplest reason is that even with a dead brain, the heart can be kept ticking! If the *heart* dies, the brain has no other option but to intone "Goodbye Columbus"! Getting back to the 'heart' of the matter, the following discussion and Figure 1 describe the uncomplicated construction of your greatest but most abused 'friend'.

The shortest and most apt description of this friend—heart is "*it's nothing but a highly efficient dual pump circulating blood through two separate closed systems*". The right side pump provides the energy necessary to move blood through the **pulmonary vessels** and the left pump provides the energy that causes flow through the **systemic organs**. Do you need a few more details? Good!

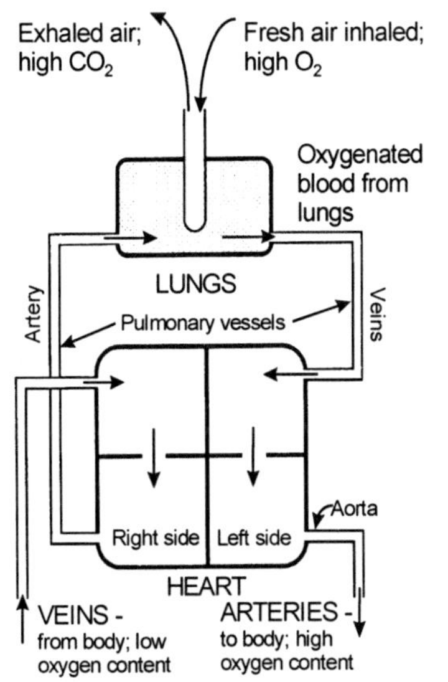

Figure 1 - Schematic of Heart & Blood Flow

Blood saturated with oxygen leaves the *left side* through the **aorta**. It circulates through the body and returns, completely depleted of oxygen, to the *right* via the **superior** and **inferior vena cavae**. The *right side* pumps this 'oxygen-less' blood through the **pulmonary artery** to the lungs, where the carbon dioxide **(CO_2)** is exchanged for oxygen **(O_2)**. The oxygen-enriched blood then returns through the **pulmonary veins** to the *left side* at which point it is ready for arterial circulation. An efficient pump it just *has* to be too, considering the fact that, in our complete circulatory system, the blood has to be pushed through more than *50,000 miles* of arteries at very precise pressure settings! Did you know that your heart also happens to be a mini-Hercules? It contracts *100,000* times daily and in the process pumps about *1900* gallons of blood per day—an absolutely sensational performance by any yardstick! In addition, it is the only part of your body that *NEVER* rests. It commenced pulsating on conception and stops beating only at death. Amazing? You bet *your* life! We'll go into the details of this life-sustaining circulation scheme in Chapter 3.

If you're gaping in awe, we're not the least surprised! So, we'll linger here and focus on a few of its more graphic aspects, without which your knowledge would remain too scanty and incomplete. No confusion or doubts so far? Great! Now for a few anatomical details of the heart.

Anatomy of the Heart

The heart is a hollow, cone-shaped muscle, about the size of your closed fist, located between the lungs and behind the lower part of the sternum (breastbone) in the centre of the chest cavity, but aligned more towards the *left* of the mid-line. That's probably why everyone grabs the left side of the chest when emoting or displaying romantic overtures!

The heart is comprised of three layers: First, the smooth inside lining called the **endocardium**; second, the middle layer of heart muscle called the **myocardium;** and third, the enclosing thin fibrous sac called the **pericardium**. A wee amount of fluid inside this sac lubricates the heart surface permitting it to move freely during contraction and relaxation.

Two-thirds of the heart is located to the left of the mid-line of the body and 1/3 is to the right. The apex (pointed end) points down and to the left. It is 5 inches (12 cm) long, 3.5 inches (8-9 cm) wide and 2.5 inches (6 cm) from front to back—roughly the size of your fist. The average weight of the heart is 9 ounces in the female and about 10.5 ounces in the male. The organ comprises less than 0.5% of the total body weight and, like any other pump, is made up of chambers and valves as shown by Figure 2.

There are basically four chambers, namely [1] the **Right Atrium (RA)**, [2]the **Right Ventricle (RV)**, [3]the **Left Atrium (LA)** and [4]the **Left Ventricle (LV)**. Each chamber has a sort of non-return or one-way valve at its exit that prevents the blood from flowing backwards. The method of pumping is that when each chamber contracts, the valve at its exit opens to allow blood to flow outwards. On completion of each contraction the valve closes and so prevents the back flow of blood into the chamber. The main valves involved are:

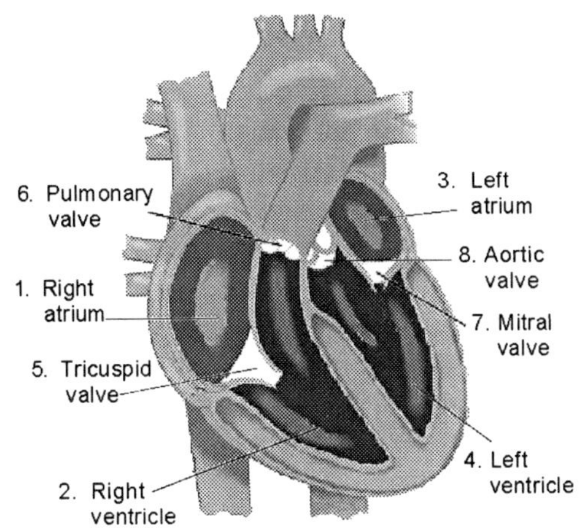

Figure 2 - Chambers & Valves of the Heart

- The **Tricuspid Valve**[5] which is located at the exit of the Right Atrium.
- The **Pulmonary Valve**[6] which is at the exit of the Right Ventricle.
- The **Mitral Valve** [7] which is at the exit of the Left Atrium.
- The **Aortic Valve** [8] located at the exit of the Left Ventricle.

The *sole function* of the heart is to *supply the required energy needed for the circulation of blood* in the cardiovascular system. Blood flow through all organs is passive and occurs only because *arterial pressure is kept higher than venous pressure* by the merry pumping of your heart. The heart's pumping action results from the innate rhythmic contraction of the heart muscles and the internal pacemaker that controls the rhythm of its beats. To maintain functional expertise there *must* be an adequate supply of blood for pumping, as well as an adequate supply of oxygen and nutrients to support the heart's muscular activity. Efficient pumping action requires the precise coordination of the contraction of millions of individual cardiac muscle cells. When everything is in perfect working order, the blood circulation performs as intended and provides the countless cells with nourishment which enables them to produce the required energy. More internal pacemaker details are discussed in Chapter 3.

Necessarily Necessary

Besides the heart, there are other vital organs in your body. Though we're not going into lengthy details regarding these, it's essential that you *know* them—and what they do—if you are to stay on track. There isn't a single square inch of your physical construction that is *not* inter-dependent in some form or the other on something else—much in the same manner as all of our organs, that despite having individual primary functions, are either supporting something else in the course of their 'duties' and at the same time sneaking *in help of their own from some other quarter.* No single part of your anatomy ever functions in isolation. That being the case, it shouldn't be too hard to conclude that, as the great Elvis Presley might have gyrated his pelvis and said, ". . . there's a whole lot of shakin' goin' on!"

It will be appropriate to do a short rundown on all the *main players* whose starring roles in this drama of life are primarily *why* we continue to live and breathe:

The Heart. Definitely the 'Team Leader'. We already have some info on him and we're certainly going to meet up with him again and again as we proceed.

The Liver. The 'big boss' of our internal detoxification plant. His job is to soften the toxic effects of the poisons and heavy metals that we are environmentally exposed to, as well as those which are natural byproducts of digestion and metabolism. His added responsibility is *detoxifying* the chemicals that we come into contact with in unhealthy surroundings. If *he* malfunctions while on the job, his primary 'dependents,' the **immune** system, the **endocrine** system and the **nervous** system will be in for a truly hard time, since all of them are strongly regulated, *and* protected from toxins, by him alone. We'll meet up with these systems a little further on.

The Kidneys. The 'CEOs' of the 'renal corporation'! Their threefold moral obligations are: 1) eliminating metabolic waste, 2) regulating the environment of the body's internal fluid and 3) producing hormones and red blood cells. But, as you'll see further on, they do very much more.

The Lungs. The 'Air Marshals' of our respiratory system. They are connected to the back of the mouth by a tube called the **trachea**, through which inhaled and exhaled air passes. While the right lung has three lobes, the left lung—with a cleft to accommodate the heart—has only two. The two branches of the trachea subdivide into smaller and smaller air vessels within the lobes. They end in tiny air sacs surrounded by capillaries. When these air sacs inflate with inhaled air, oxygen mixes into the blood in the capillaries, which is then pumped by the heart to the various tissues of the body. While oxygen is mixing into the blood, at the same time, carbon dioxide disperses out of the blood into the lungs, from where it is then exhaled.

The Immune System. This system has the responsibility of protecting the body from disease-causing organisms such as bacteria, virus, fungus and yeast. In addition, it plays a significant role in controlling how reactive the body is to food, pollen, dust and other agents that cause allergy. It also determines how easily the body develops an inflammatory response. If this system is weak, then your body is going to be at the receiving end of a *chronic-disease* attack!

The Endocrine System. This 'conglomerate' is responsible for health maintenance purposes. The functions of its various subsidiaries are as follows:

> ***Adrenal Gland*** activity is important for maintaining energy throughout the day and for responding to stress, whether physical or psychological.
>
> ***Thyroid Gland*** activity is important in maintaining proper metabolism and is also important for maintaining energy and proper body weight.
>
> ***Testes*** and ***Ovarie***s have functions that are significant for production of male and female hormones which maintain health for sexual functions, help maintain proper body weight and energy, and in addition, also influence moods.

The Pituitary Gland ensures that this entire system is working together in unison.

Melatonin activity is important for sleep and mood besides regulating body temperature. It also affects reproductive health.

Pancreas functions are important in maintaining proper blood sugar levels.

Thymus Gland activity plays a significant role in maintaining the health of the immune system.

The Nervous System is responsible for taking in information through the functioning of the ***five senses*** and helping us to meaningfully interpret this input. It allows us to be sensitive to the world around us through nerves that disseminate the sensations of touch, temperature and pain throughout our system. It controls muscle tone and muscle activity. It also has significant **automatic regulatory control** over most bodily functions including digestion, heart rate, the size of our breathing passages, the rate of breathing and the size of the contractual abilities of the eyes' pupils. Our higher functions such as memory, learning, the ability to manifest creativity and the expression of moods and feelings, remain largely under the direction of this system.

Skirting the Red Issue!

Hang in there—we'll 'doctor' your dispositions yet! Having 'disposed' of some of the main components that keep your body ticking, we have arrived at that important juncture where descriptive aspects have to be more detailed. Okay, okay—so we'd promised **NO** medical tome; we're sticking to that! But we're talking here about life's most marvelous creation—the *human body*. Since most of us know *absolutely nothing* about it and understand it even less, we shouldn't be surprised at the fact that, more often than not, we grossly mistreat our *very own selves*! There is also another problem with human structure; when pinning the adjective 'important' it is very difficult to actually categorize the degree, simply because, to your chagrin, *and* ours, every part and function of our system happens to be equally IMPORTANT! In this context, the very reason that you saw ***red*** a moment ago is just what we're going to expound upon–your blood!

The Fluid of Life

Your blood, or the fluid substance that circulates within you, is a vital liquid that performs many critical functions. It nourishes every tissue, every gland and every organ. It not only transports oxygen and nutrients to the cells, but also acts in removing impurities and waste products and transporting these for detoxification and elimination from the body. Blood circulation takes place in the vast labyrinth of arteries, veins and tiny vessels called capillaries. Its colour is bright red or scarlet when oxygenated, and it becomes bluish-red when the oxygen supply has been depleted. Blood is composed of water and **plasma**. Plasma is a yellowish fluid in which millions of cells are suspended. It has a characteristic odour and constitutes about 45 percent of the blood volume. The remaining 55 percent is water; the medium that facilitates the circulation of the many indispensable factors of which blood is composed. A cubic millimeter of blood contains about *5 million* **red corpuscles** called **erythrocytes**, *5000 to 10,000* **white corpuscles** called **leukocytes** and *200,000 to 300,000 platelets* called **thrombocytes**. It also carries many salts and organic substances that are in solution. The average, healthy adult body carries between 4.5 to 6 litres of blood. In getting to know the 'whys' and 'wherefores' of the various active components, we've unwittingly revealed where the term *a bloody nuisance* originated! Nevertheless, here goes:

Erythrocytes. These are the red corpuscles, or red blood cells (RBCs), of which the mature ones do not contain a nucleus. These red cells are formed in the bone marrow and after an average life of 120 days, during which time they incur substantial damage, they are broken down and removed by the **spleen**. There should be between 4.5 and 5.5 million red blood cells in the body. An *increased* number denotes inefficient breakdown or improper production. *Decreased* numbers indicate inefficient production or possible deficiencies. **Hemoglobin** is the blue-red, iron-containing compound residing in these cells and is reflective of the amount of intercellular iron. Increased hemoglobin indicates a hyper-active spleen while a decreased percentage indicates an anemic condition, usually resulting from deficiencies.

Leukocytes. These are the white corpuscles, or white blood cells (WBCs), which are reflective of the body's defence mechanism and work as an "infection control indicator". There are between 5000 and 6000 of these cells. Increased numbers of cells indicate the possibility of tissue destruction and/or an infection within the body. Decreased numbers indicate poor cell production and susceptibility to disease. Leukocytes exist in two principal forms or types:

a) The granular type, formed in the bone marrow, with multi-lobed nuclei that contain:

Neutrophils which ingest and destroy bacteria.

Eosinophils which increase and become active in the presence of certain infections and allergies.

Basophils which secrete **heparin** and **histamine**. Heparin is the anticoagulant while histamine stimulates inflammation. An increase of these substances reflects a possibility of parasitic activity in the body.

Monocytes normally constitute 3% to 7% of the circulating leukocytes. An increase of these may point to inefficient liver function.

b) The non-granular type with rounded nuclei are the **lymphocytes** which are formed in the **thymus**, the **lymph** glands and other lymphatic tissue. They are associated with the immune phenomena since they are not only elaborate antibodies, but are also concerned with cellular immunity. An increase in these substances is indicative of excessive toxin production in the body. Decreased numbers of lymphocytes indicate possible active body infection.

Thrombocytes. These blood platelets—also formed in the bone marrow—are small, rounded non-nucleated bodies. They adhere to the site of an injury and thus plug the defect in the vascular wall. As they disintegrate, they release coagulating agents leading to the local formation of **thromboplastin,** which helps to form a clot—the first step in the healing of an injury.

Plasma. This is a very complex substance. It contains proteins, many inorganic constituents, sugar, hormones, enzymes, fats, amino acids and waste products such as **urea** and **creatinine**—all of which occur in minute quantities. The proteins also act as carriers of essential nutrients such as copper, iron, other metals and a variety of hormones.

All the constituents of the blood are continuously being worn out or consumed and have to be repeatedly replaced. These components are formed or generated in various organs of the body including the liver. For instance, the endocrine glands make the hormones that are transported in the plasma; the lymphocytes and plasma cells make some of the proteins and some other constituents are derived by absorption from the intestinal tracts.

Circumventing the Circulation

By now you must have definitely appreciated exactly why blood is termed "*the fluid of life*". Since it's the life-sustainer of every micro inch of your anatomy, it naturally follows that the blood *must* possess a very complex, exhaustive and extensive circulatory system. No need to slap your forehead in alarm; these are essential descriptions! It wouldn't be too far off the mark to consider your blood as a sort of 'carrier-cum-transporter'—it's the only item inside you that's constantly on the move! Since no part of you is ever left untouched by blood flow, it is also true that the impurities and waste matter which are removed for final eradication from the system sometimes get stuck en route and create a kind of biological imbalance at certain junctures. To enable an easier grasp of the intricacies of blood circulation, we'll take the simple headache as an example. When you swallow an oral painkiller or just apply some balm to the site, the resulting process of both actions is the same. The tablet dissolves in your stomach and from there *the blood* transports the sedation agents to the right nerves. In the second instance, the blood absorbs the balm through the pores of the skin and spreads it around through the capillaries and tiny extremity arteries. This explanation may sound like an oversimplification, but—in essence—that's exactly what happens.

"Are you telling me that my blood can carry germs as well?" you might interject. Absolutely correct! There are simply no other routes that germs, viruses or bacteria could possibly follow *but* via these pre-designed and existing channels for blood circulation. Is the light beginning to dawn? You're right on target if you conclude that not only does your blood keep your 'engine' running, but it also is the principle 'engine-seizure' respondent! That is why the diagnosis and reasons, for ***every*** ailment or physical problem, are only invariably available *after* an analysis of the blood! Don't make the mistake of assuming that your blood is the main villain of the piece. Not so! Your blood is the actual champion; it is *you and me* who are the real culprits. In this day and age, with new diseases cropping up faster than cures for the already existing maladies are found, the ultimate onus for maintaining and preserving health must ***perforce lie with the individual.*** We can no longer rely on that convoluted truism that "for every medical problem, there is a doctor available". Therefore, it becomes incumbent upon the individual to study and learn about the ways and means of health for himself. So please curb that urge to bite our heads off every time we start spouting some 'medicalese'! Now you know why!

Isn't Chemistry for the Labs?

Definitely true. But don't overlook the fact that inside of you is the most complex, most efficient and most remarkable 'laboratory' that ever was, or ever will be! It is at work non-stop, even when you sleep, and the variety and number of ongoing 'experiments' is truly mind-boggling. If any man-made labs ever tried to emulate such complexities or working conditions, more than just a few gaskets would blow through the roof! Hence, at this point, it is imperative that you understand what a balanced blood chemistry means, what nutritional needs are and, to further expand your knowledge, the implications of the ways and means of health. We have been led for far too long to believe that 'normal' means 'healthy'. Many doctors equate these terms on the same level when they consider a *normal* or *healthy* state to mean the *absence of illness or disease.* In the context of blood tests, the normal ranges of the results are an indication of whether the individual is free from any disease or not. However, these results may not indicate what's actually happening or breaking down at the metabolic level—where the signs and symptoms are so slight that they may go unnoticed. It is only after pinpointing the actual malfunction and eradicating it that biological healing can really take place. In simple terms, if an unbalanced blood chemistry can be corrected at the very first instance by itself, then the body's internal environment will always be able to retain its stability. In bionutritional analysis, blood chemistry values are interpreted within a narrower range than does conventional medicine or pathological interpretation. This narrower range is called the **sub-clinical** range. A sub-clinical blood chemistry value is one that is outside the optimum range but does not reach the pathological level. These deviations in blood chemistry are related to physiological, metabolic and biochemical malfunctions. These malfunctions have been found to indicate specific nutritional deficiencies that can be corrected by appropriate nutritional supplements.

For example, the standard accepted range for serum calcium is 8.5-10.5 mg per dl. (decilitre). Values above 10.5 can indicate a pathological condition such as **multiple myeloma**. A value below 8.5 can indicate **rickets** or **osteoporosis**. No significance is given to the values between 8.5-10.5. ***However, according to us the optimum range for calcium is 9.7-10.2 mg per dl. Values above 10.2 or below 9.7 can indicate digestive malfunctioning, which can be corrected through proper diet and supplementation.*** If the blood chemistry can be corrected or balanced in this way, it is possible to prevent the development of pathology, as well as improving overall health and energy. *This technique can be applied to all of the 33 characteristics of blood chemistry.* It is also possible that pre-arthritic, pre-cancerous and pre-thyroid conditions may be detected in the body several years in advance of their development and can be avoided/prevented with a comprehensive programme of nutrition supplementation.

See? There's really no need to get 'chemical' to remain physically intact! Now for a peek at some of the component elements of blood chemistry, what they mean and what sort of influences they exert:

Creatinine is a waste product derived from the muscle mass and is normally removed almost entirely from the body by the kidneys. The creatinine reading reflects on skeletal muscle anabolism (repair and healing ability) and catabolism (destruction of cells) in the body maintenance plus an indication of kidney function efficiency.

BUN (Blood Urea Nitrogen) is produced by the liver from a breakdown or digestion of protein and is excreted by the kidneys. It tends to elevate above normal if there is an excess protein intake or a decrease of kidney excretion. Increased BUN points to **liver** and/or **thyroid** inactivity. Decreased BUN indicates **pancreas** and/or **adrenal** inactivity.

BUN/Creatinine Ratio shows the body's intelligent perspiration ability. An increased ratio indicates that a person does not sweat, or sweat readily, but may have toxic night sweats. A decreased ratio indicates the tendency to perspire more readily.

Calcium is important for both bones and teeth. It carries protein and fat across the intestinal membrane after digestion and cooperates with ***phosphorous*** in many bodily reactions. A high percentage of calcium indicates poor breakdown of fats, poor gall bladder function or perhaps improper protein digestion in the liver. A low percentage indicates poor chemical changes of fatty acids plus poor protein digestion.

Phosphorous often works in tandem with calcium and is important in human physiology. It is a constituent of all bones in the form of **calcium phosphate** and carries carbohydrates to the liver for complete digestion, storage and utilisation. Increased levels of phosphorous indicate deficient hydrochloric acid in the stomach which impairs digestion of all proteins causing burping, flatulence, constipation, dry skin and eczema. A decreased level indicates an over-acid condition in the stomach with probable impairment of protein digestion resulting in heartburn, hunger, obesity, storage of water and fat, oily skin and acne.

Sodium reflects body fluid control and kidney function. Its increase indicates an alkaline kidney and lack of membrane lubrication often resulting in fluid retention as well. Decreased sodium indicates an acid kidney membrane plus the need for calcium—and sodium—rich foods in the daily dietary intake.

Potassium is a mineral normally found in high concentrations inside body cells and is necessary for normal nerve transmission plus other functions. It reflects tissue composition of the muscles, especially the heart muscle, and the mineral exchange across the **sinoauricular node** or the heart's pacemaker. Increased potassium indicates tissue composition with congestion possible within the heart muscle especially when combined with an unusually slow heartbeat. Decreased potassium indicates hyperoxidation with a somewhat weakened heart muscle, especially if combined with a rapid heart beat.

Chloride presence affects and reflects the proper function of the fluid exchange membranes (bowel and bladder). Increased chloride indicates improper membrane lubrication by a vitamin. Its decrease may indicate tissue breakdown in the bowel and/or bladder and, perhaps, even in other parts of the body.

Uric Acid is also a product of protein breakdown, in fact, it is the end-product of protein resolution and is normally secreted by the kidneys. Increased uric acid indicates incomplete and/or impaired protein metabolic management, and/or poor elimination via the kidneys (resulting in gout or renal dysfunction). Decreased percentages indicate incomplete protein digestion, and/or a deficiency of protein in dietary intake, and/or an over-acidic condition of the kidneys with possible inflammation of these organs and dryness of joints (arthritic syndrome).

SGOT (Serum Glutamic Oxalacetic Transaminase) AST indicates and reflects gonadal function and also shows the amount of oxygen available at a membrane. Increased SGOT indicates a deficiency of hormones and Vitamin E, while a decreased percentage indicates deficiency of the gonad itself.

SGPT (Serum Glutamic Pyruvic Transaminase) ALT reflects the function of the liver. Increased SGPT shows a need for vitamin lubrication of Vitamins A & D while a decreased reading indicates the need for copper and iron in the liver and spleen support.

Albumin is the **colloidal** (substance in a non-crystallized state with very large molecules) protein synthesized by the liver that controls osmotic (tendency to diffuse) pressure of fluids to promote transfer of nutrients and wastes to and from the blood and cells. Increased albumin indicates high blood viscosity (thick blood), which could be due to improper protein digestion or dehydration. Decreased albumin points to low viscosity of blood with possible water retention, nutritional congestion, toxic build up and edema. It is also evident in cases of gross starvation.

Bilirubin is formulated in the liver and eliminated by it to the intestinal tract. This red bile pigment is found as **sodium bilirubinate**, which is a soluble compound, or as an insoluble calcium salt in gallstones. It influences the functions of the lymph and spleen systems. Increased total bilirubin indicates inefficient lymphatic or liver-gallbladder function or that lymphocytes may have deviated from the norm. A decreased percentage indicates inefficient blood cell breakdown and perhaps also inefficient function on the part of the spleen.

Globulin reflects the amount of protein involved in body defence mechanisms or, the amount of protein in the body's immunological defences. Too high a percentage of globulin indicates chronic infection or increased amounts of a non-specific protein. Decreased globulin points to impaired protein digestion and metabolism, malnutrition or an altered gastrointestinal tract absorption plus a chronically depressed defence system.

Total Protein reflects all the proteins available for use by the body for purposes of healing and repair, for pancreatic function, pH (acidity or alkalinity) maintenance, muscle metabolism and for emergency supplies of fuel. It is a guide to pancreatic function, the pH of the stomach and general muscle metabolic efficiency. This is not necessarily connected to dietary intake and should be considered in relation to albumin, globulin and also creatinine.

A/G Ratio is the ratio between albumin and globulin. An increased ration indicates increased viscosity of the blood and tendency for clots. A decreased reading indicates poor protein digestion.

Alkaline Phosphatase reflects the pH of the blood which is normally slightly alkaline. It also reflects the effect of the blood pH on adrenal physiology and posterior pituitary. An increased reading indicates an alkaline blood system (too alkaline) with insufficient mineral transfer to cells; however, an elevated reading can be entirely normal in children. If both LDH and alkaline phosphatase are elevated together, a primary liver condition exists. A decreased reading indicates an exhausted adrenal system and an acid blood system. If both LDH and alkaline phosphatase decreased together, a primary thyroid deficiency should be suspected.

Glucose values are used to assess the sugar in the blood. Elevated values after a 12-hour fast are evident in diabetes. Increased glucose may indicate a dietary overload with starch and/or fat foods; also, a lack of suitable enzymes, catalysts and/or hormones to regulate metabolic acceptance. Decreased glucose may indicate the need for complex carbohydrates in the dietary intake. It may also indicate an overload of enzyme catalysts and/or hormones to regulate metabolic management.

Cholesterol is the fatty portion of protein, which supplies enzymes, hormones and antibodies. An increase could mean an excessive amount of rancid acids in the blood stream or poor distribution of cholesterol due to hormonal imbalances. A decreased reading could mean a deficiency of fat-soluble vitamins and/or a sluggish liver and thyroid, or an overactive production of hormones by certain glands. Total cholesterol is made up of **high density lipoprotein** (HDL) and

low density lipoprotein (LDL). HDL may be termed as the 'good' cholesterol and higher values indicate more efficient waste matter clearance. LDL tends to deposit more waste in the arteries and a high value is indicative of increased clogging. Abnormalities of lipid metabolism result in small dense particles of LDL called **apolipoprotein(a)** [Apo(a)] which appear in elevated levels when increased coronary atherosclerosis is evident. **Apolipoprotein(b)** is similarly produced but is more closely associated with ***VLDL*** (very low density lipoprotein) which is a mutated form of LDL.

Triglycerides reflect the body's balance of the nervous system metabolism. This fat is normally present in the body and is broken up into fatty acids which can be used as fuel. Increased tryglycerides indicate a tendency for slow utilization with mineral buildup plus a depressed nervous function. A high level of triglycerides present over a long period of time is associated with the premature hardening of the arteries. The most common cause of an excess in the blood is over-eating. Alcohol, sugar, birth control pills or diabetes can further elevate this fat. When the level decreases, it indicates a tendency for rapid utilization of nutrients with mineral deficiency and also an increased stress sensitivity to stressors.

Lipoprotein(a) also simply known as Lp[a], is a genetic factor closely related to LDL but is a more deadly variety. Lp[a] levels present soon after birth and serum levels are shown to correlate well with the presence, extent, severity and lesion score on coronary angiography as well as with the occurrence and recurrence of myocardial infarction and cardiac death. High serum levels are also associated with bypass graft stenosis, valvular atherosclerosis and persistently occluded coronary arteries.

Homocysteine is an amino acid that is a by product of protein metabolism and results from a genetic mutation. It is a major cause of atherosclerosis and even-affects the arteries feeding the heart and brain. High levels promote blood clots in the veins as well as in the arteries and are indicative of the body's deficiencies of folic acid and other B-complex vitamins. Homocysteine levels tend to increase with age.

Lactic Dehydrogenase (LDH) is an intercellular enzyme present in the heart, liver and skeletal muscles, which reflects blood acidity and pancreatic functional efficiency. It also assists biological oxidation (balances water and sugar). Increased LDH indicates diabetic type tendencies with swelling of acid tissues and decreased circulation. Decreased LDH often leads to a condition of extreme fatigue besides indicating pancreatic hypoglycemic type tendencies and an alkaline tendency of the blood.

Carbon Dioxide (CO_2) reflects lung condition with O_2 and CO_2 exchange for proper acid/base balance. Increased CO_2 indicates a need for proper vitamin lubricators & possible lung congestion with a mucous-productive cough. Decreased CO_2 indicates dry lungs, low vitamin A lubrication with tissue destruction plus a dry, non-productive cough.

Iron influences activity of the **spleen** as well as the red blood cell production by the bone marrow. Increased iron usually reflects increased bone cell production and a possible need for more RBCs and WBCs besides emphasizing the greater need for exercise and exposure to sunlight.

T-4 (Tetraiodothyronine) (THYROXINE) reflects thyroid function-related basal metabolic rate. A high T-4 shows a hyperactive thyroid or compensation for other stresses or weaknesses related to adrenal, pituitary, hypothalamus or thymus. Decreased readings indicate a hypo-active thyroid with impaired activity of basal metabolic rate, liver and adrenals.

Hematocrit reflects the total amount of solids in blood, mostly blood cells. It is the percentage of whole blood which is composed of red blood cells. An increased percentage indicates possible hyperactive spleen, active infection or dehydration. Decreased hematocrit indicates possible low blood cell production or water retention

SEGS (Segmentals) (Neutrophils) (Polys). Segmentals are bits and pieces of parts and organs, essentially protein matter which, for various reasons, have been flayed off by both normal and abnormal processes of the body. Increased SEGS indicate excess foreign protein or active infection. Decreased SEGS point to the possibility of excessive toxin production in the body.

Metal Traces in My Body? Nuts!

No, we're not crazy. There *are* metallic bits and pieces floating around inside you—so tiny that you'd need the most powerful microscope to get even a trace. These are medically termed as **trace metals**. It is important to keep in mind that if these trace metals do exist, then they're hardly likely to be sitting around having a cup of tea! They will be milling around like a bull in a china shop, generally causing havoc, and the only way that they can get from one place to the next is by being carried by the blood! If some of them are unwanted or if the needed ones exist in excess of what is actually required, then you're definitely in for some unnatural bodily contortions! We shall now take a look at some of the biological and toxicological properties of these trace metals.

Aluminum. There is no known biological requirement for this metal in the human system, but small amounts are normally present from the food we eat and the dust that we inhale. Excess exposure in the body can produce pathological changes in the brain, lungs and bones causing inhibiting infirmities in some vital organ and could also be a contributing factor in causing Alzheimer's disease.

Arsenic. Chronic intoxication of this metalloid is caused by exposure to arsenic pesticides or herbicides or through the consumption of contaminated seafood such as shrimp and lobster. Acute/severe arsenic poisoning results in the paralytic symptoms of shock, vasodilation, cyanosis, coma and death from circulatory failure. The pesticides cause skin, lung and lymphatic cancers. This material is largely secreted through the urine as methylated derivatives.

Cadmium. This is ingested through industrial exposures such as steel production as well as pigment and battery manufacture. Dietary sources such as rice, wheat, oysters and other sea food as well as smoking contribute to the body's burden. Acute airborne exposure can bring about respiratory distress, while chronic exposure results in kidney tubular damage.

Chromium. The commercial uses of this metal, such as steel manufacture and metal plating, are well known. As far as the body is concerned, trace amounts are considered essential in glucose and lipid metabolism. Chromium is necessary for normal glucose utilization, complexes containing proteins and the stimulation of several enzyme systems. It may stabilize certain nucleic acid structures and also stimulates synthesis of fatty acids and cholesterol in the liver. The **hexavalent** form of chromium is known to be mutagenic and carcinogenic while the **trivalent** form is considered safe in physiological amounts.

Copper. As an essential trace metal, copper is an active component of several proteins and important enzymes. It must be present to convert iron into hemoglobin. It influences tyrosine, an amino acid, and also utilizes Vitamin C. Determination of copper in biological samples is used to diagnose metabolic diseases due to deficiencies or excessive accumulation.

Iron. An essential metal which is maintained under close homeostatic balance for hemoglobin production and other biological needs such as the body's defence against infection. Deficiencies of iron can result in anemia with a decreased capacity for oxygen transport and energy. Genetic iron overload is detected by measuring iron levels in liver biopsies.

Lead. This metal is particularly hazardous to children and mothers of child-bearing age. Exposure can occur through the paint in old houses, pottery glaze, lead batteries and the lead solder in water pipes. It can also be absorbed through the atmosphere when breathing air polluted by exhaust gas from vehicles using leaded fuels. Development of mental retardation, anemia and encephalopathy are a few of the clinical effects of lead toxicity.

Magnesium. This metal serves as a starter for some of the chemical reactions in the body. It plays an important role as a coenzyme in the building of protein and as a sleep-promoting material. A deficiency relative to calcium may lead to accumulation of calcium salts in the kidney serum and in the arterial system resulting in **renal calculi** and atherosclerotic plaques. Systems affected by acute deficiency are primarily neuromuscular, cardiovascular and renal. Initial manifestations are hyperirritability leading to convulsions, hyperemia and vasodilation.

Manganese. This works with the B-complex vitamins to overcome laziness, sterility and marital weakness. It combines with the enzyme **phosphatase** to build strong bones and is also needed for good enzymatic function so that foods are digested. It helps build resistance to ailments, enhances nerve health and promotes milk production in expectant mothers. It is also involved in pigment formation.

Mercury. Exposure to this highly volatile metal can occur from the air but the main sources are from contaminated seafood, industrial wastes and fungicides. Degeneration from overexposure presents itself as a sensory nervous disturbance in the brain and/or proximal kidney tubular damage.

Selenium. An essential component of **glutathione peroxidase**, selenium plays a role as an antioxidant. Dietary deficiency has resulted in cardiomyopathy, an impaired immune response and the onset of cancer. Toxicity from overdoses results in several metabolic and physical changes such as impaired bone development, liver disease, hair loss and fatigue.

Silicon. The role of silicon is not well defined but it functions as a structural component of membranes in connective tissue. The possible protective role of this element appears in the increased amounts found in **hemodialysis** patients and as **aluminosilicate** in the brains of Alzheimer's patients. Its toxic properties are indicated if local inflammations occur and as scleroderma in women with silicon-gel breast implants.

Thallium. This was earlier used as a fungicide, insecticide and rodenticide because it was highly toxic. It had also been used as a venereal disease treatment but currently it is used only as a tracer in myocardial imaging. Environmental and occupational exposure can be detected through the urine.

Zinc. This is a constituent of both insulin and the male reproductive fluid. It combines with phosphorous to aid in respiration and sparks vitamin activity. It helps food become absorbed through the intestinal wall and is also essential for the synthesis of collagen.

You should now fully appreciate that your blood is much more than simply a 'run of the mill' liquid—in fact, it is the most complex and versatile fluid ever created. With so many components and constituents, each having equal degrees of importance, is it any wonder that the tiniest of imbalances or dysfunctions—even in microscopic measurements—can upset the whole apple cart? Contrary to any of the impressions that you may have acquired while reading through this descriptive gamut, we are not advocating a means of self diagnosis or self treatment, but rather are providing you with the means to equip yourself with the education and knowledge you need to maximize your own health, something for which only **YOU** can be responsible!

CHAPTER 3

MANAGING THE CIRCULATORY SYSTEM

A Thoroughly 'Heartfelt' Squeeze

It is once again time to confront the 'Team Leader' (your heart) and to understand ***exactly*** how he manages to control and manipulate the circulatory system with such efficient dexterity and how he also handles a series of sub-systems in the course of this task including the heart beat 'Command & Control' system. And a heartfelt squeeze it is! An average healthy and normal heart can manage to pump – squeeze – approximately 5 litres in the 72 times that it beats every minute. In other words that works out to 7,200 litres (about 1900 gallons) per day, almost 2,628,000 litres per year and an astounding 184,086,000 litres by the age of 70! Not too bad for a 750 gram flesh pump that's probably the most mistreated organ in the body! And we cannot overlook the Team Leader's durability; The average heart beats 72 times per minute which in the course of a single day amounts to over 100,000 times. Calculating further, we find that in a year, the heart beats almost 38 million times, and by the time you attain the ripe old age of 70, on the average, *it has completed 2.5 billion beats!*

So, What Is the Pipeline?

As mentioned earlier, the 1,900 gallons of blood that the heart pumps daily flows through about 50,000 miles of blood vessels, arteries, veins and capillaries. Since our body's construction is so perfect and precise, the fabrication and routing of the 'plumbing system' is naturally nothing less than superlative. From the tips of your extremities to the cores of your vitals, every atom is fed, cleansed, detoxified, refurbished, energized, vitalized, regenerated and purified ***by your blood***. If this be the case, then that's all the more reason for you to completely understand the make up, set up and working of your blood's 'conduits'. The importance of *this* knowledge will gradually manifest itself as we continue. Some basic explanations appear below:

The **Circulatory System** of the human body is the course taken by the blood, starting from the heart, through the arteries, capillaries and veins and then *back* to the heart.

Arteries are the pipelines carrying oxygen-enriched blood from the heart to the capillaries for further distribution throughout the system. An artery is made up of three layers—a tough outer layer to withstand blood pressure, a muscular middle layer which allows for expansion and contraction with the pulse and helps propel the blood along and, lastly, a very sensitive inner lining. It is imperative that the arteries remain healthy and fully functional if the body is to perform all its tasks efficiently.

Capillaries are the tinier conduits to every nook and corner. Taking the oxygenated blood from the arteries, they distribute this rich 'food' to the required sites and after the 'banquet', transfer the oxygen-depleted blood to the veins for the return journey back to the heart. In the flow through the capillaries, the red blood cells move along rapidly in the middle of the current while the white cells advance more slowly on the periphery along the walls. Capillaries present a very large surface, greater than any other blood vessel, for the blood to come into contact with. They offer the greatest resistance to the flow of the blood and, as a result, have a great influence on the total circulation. These pipelines expand with the rise in temperature and help keep the blood cool. When it is cold, they contract and so help to preserve internal heat.

Veins are the vessels that conduct deoxygenated blood from the capillaries back to the heart. There are three exceptions to this description. 1) The **pulmonary veins** return blood from the lungs, where it has been oxygenated, to the heart; 2) the **portal veins** receive blood from the pyloric, gastric, cystic, superior mesenteric and splenic veins and, on entering the liver, break up into small branches that pass through all parts of it; 3) the **umbilical veins** convey blood from the fetus to the mother's placenta. Veins tend to enlarge as they proceed, gathering blood from their tributaries and finally pour this collected blood into the right atrium of the heart. The coating of veins is similar to those of arteries, but is thinner and often transparent.

Flow of Blood

As an overview, the circulatory system consists of the following four subsystems:

1. **Pulmonary Circulation.** This subsystem makes use of the *right side* of the heart to pump oxygen-depleted blood, returning from the body, to the lungs through the **pulmonary artery.** During the bloods' passage through the lungs, carbon dioxide gas is removed and it absorbs oxygen. The oxygenated blood is then transported back to the heart through the four **pulmonary veins** and enters the left atrium for further pumping by the left side of the heart to the rest of the body so that every single cell in the body will be supplied with much needed oxygen.

2. **Body Circulation—Branching.** The **aorta**, the heart's discharge vessel, divides into a number of main arterial branches which, in turn, divide into smaller ones until the entire body is supplied by an elaborately branched series of blood vessels. The smallest arteries, **arterioles**, divide further into a fine network of still more minute vessels, the **capillaries**, that have extremely thin walls which enable the blood to come into close contact with other fluids and tissues of the body. The three functions of the blood in the capillaries are 1) to release oxygen to the tissues, 2) to furnish the cells with nutrients and 3) to carry other essential substances such as the waste products collected from the tissues. The capillaries then unite to form veins which further unite with each other to form larger veins, until the blood is finally collected in two large veins, the **superior vena cava** and the **inferior vena cava**, and returned to the heart, thereby completing the circuit.

3. **Portal Circulation**. This is a subsidiary to the venous system and exists in addition to the pulmonary and systemic circulation. A certain amount of blood from the intestine is collected into the portal vein and carried to the liver where it enters into open spaces called **sinusoids** and comes into contact with liver cells. Within the liver, important changes occur to this blood which is carrying the products of digested food recently absorbed through the intestinal capillaries. This blood is collected a second time into the veins where it rejoins the general circulation through the right atrium. Blood is also subjected to further modifications in its passage through other organs.

4. Coronary Circulation. This function is the one exception to the rule that blood has to pass from the right side of the heart to the left ***only*** by way of pulmonary circulation through the lungs. This circulation, also being free of wastes, supplies the very heart tissues themselves with nutrients and oxygen. As the blood is pumped out of the heart and into the aorta, a portion is routed back to the heart via the coronary arteries—three separate vessels that travel across the surface of the heart, dividing, branching and ultimately penetrating the heart to supply it with oxygen and nutrients. Blood from this coronary capillary circulation enters several small veins, which then enter directly into the right atrium ***without*** first passing through the vena cava.

Now we'll combine the first three of the above subsystems and follow the blood's path through the body.

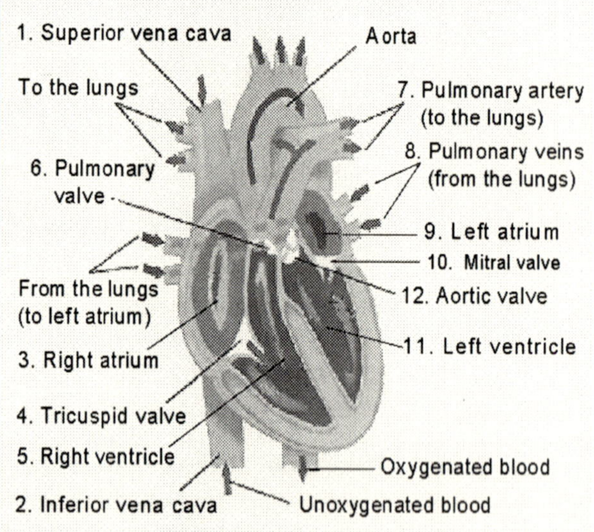

Figure 3 - Flow Paths and Components of the Heart

All blood enters the right side of the heart through two veins, the **superior vena cava (SVC)**[1] and the **inferior vena cava (IVC)**[2] (see Figure 3). The SVC collects blood from the upper half of the body while the IVC rounds up the fluid from the lower half. On leaving the SVC and IVC the blood enters the **right atrium** [3] and when the RA contracts, it transits through the **tricuspid valve** [4] into the **right ventricle** [5]. When the RV contracts, the tricuspid valve closes and the blood is pumped through the **pulmonary valve** [6] into the **pulmonary artery (PA)** [7] and on further into the lungs where the oxygen-depleted blood picks up oxygen. So it is the ***right side of the heart,*** which first pumps the blood to the lungs for a pick up of fresh oxygen after which the blood goes to the ***left side of the heart.*** Now, recharged with oxygen, it is pumped out and returned to the body.

The blood returns to the heart from the lungs by way of the **pulmonary veins** [8] and enters the **left atrium** [9]. On contraction of the LA, it pushes through the **mitral valve** [10] and enters the left ventricle [11]. The LV is a very important chamber in that it is responsible for pumping blood through the **aortic valve** [12]) into the **aorta** [13] and finally to the rest of the body. The ***aorta is the main artery of the body*** as it receives all the blood that the heart has pumped out and further distributes it to the rest of the body. The LV has a thicker muscle than any of the other heart chambers because it is required to pump the blood to the rest of the body *against much higher pressure* in the general circulation system (blood pressure).

For a simple memorization of the flow, the adjacent flowchart 'shows' the way. For more details of the circuit through the body, the Figure 4 flow schematic

Oxygen-depleted blood from the body returns to
⇩
the *Superior* **and** *Inferior Vena Cava* **and then**
⇩
to the *Right Atrium* **and**
⇩
through the *Tricuspid Valve*
⇩
to the *Right Ventricle* **for a further push**
⇩
through the *Pulmonary Valve*
⇩
to the *Pulmonary Artery* **and on**
⇩
to the *Lungs* **where oxygenation takes place, then flows out**
⇩
through the *Pulmonary Veins*
⇩
to the *Left Atrium* **and**
⇩
through *Mitral Valve*
⇩
to the *Left Ventricle* **for a high pressure push**
⇩
through the *Aortic Valve*
⇩
to the *Aorta* **for distribution**
⇩
to the rest of the body
⇩
return to the top and repeat the cycle

shows how much of your blood goes where. The flow splits through the body are typical for an individual at rest.

The Heart's Blood Supply

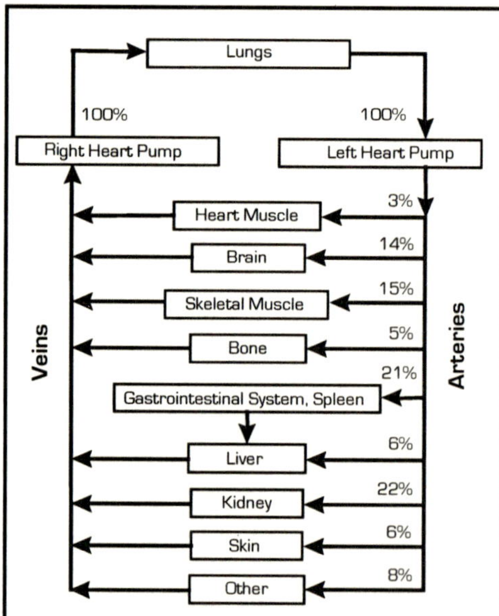

Figure 4 - Cardiovascular Circuit & Flow Division for an Individual at Rest

We've examined how the first three subsystems supply blood throughout the body. Now let's examine how the fourth subsystem, the coronary circulation, supplies blood to the heart.

The **coronary arteries** are the ones that we *MUST* keep 'clean'. This can be achieved by resorting to a regulated healthy diet. If these arteries somehow get 'blocked', you can ***rest assured*** that a heart attack ***will*** result! Just like all other organs, the heart also requires the blood to supply it with oxygen and other nutrients so that it is able to continue working efficiently. *The heart **does not** extract oxygen and other nutrients from the blood **flowing inside it.*** The heart gets its own oxygen-rich blood supply from coronary arteries that eventually carry the blood into the heart muscle. Approximately 3 to 5 percent of the blood output of the heart goes to the two main coronary arteries (see Figure 5)—the **left main coronary artery** [1] and the **right coronary artery** [2] both of which arise from the aorta. The left main coronary artery further divides into the **left anterior descending** [3] and

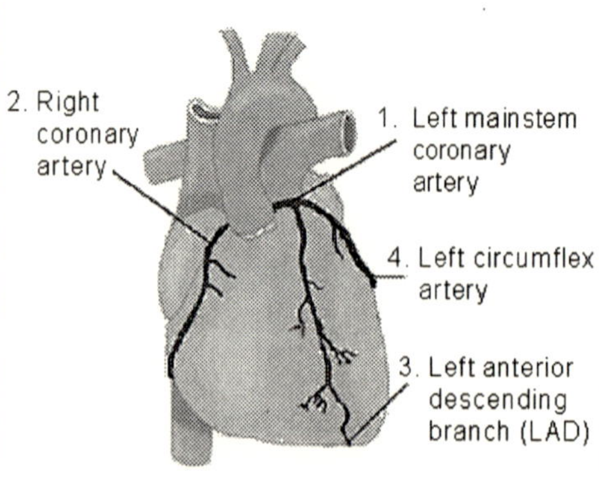

Figure 5 - Blood Supply for the Heart

the **left circumflex** [4] arteries. Each artery supplies blood to different parts of the heart muscle and the heartbeat electrical system.

The heart also contains the veins that collect oxygen-poor blood from the heart muscle. Most of the major veins of the heart—*great cardiac vein, small cardiac vein, middle cardiac vein, posterior vein of the LV, and oblique vein of the LA*—drain into the **coronary sinus** which opens into the RA. Coronary artery disease is basically caused by any blockage in one of the coronary arteries. When a coronary artery is partially blocked, that particular artery just cannot supply sufficient blood to meet the essential needs of the heart muscle during exertion. Anyone with coronary artery disease will definitely experience chest pains during exertion because of a lack of blood and oxygen to that particular part of the heart muscle. This pain is medically termed as **angina.** This symptom of coronary artery disease will be encountered several times in following chapters with the most detailed discussion being in Chapter 10.

Boogie to the Beat!

At this 'intake' rate, you can take a wild stab at the next Medical Certification Board exams! And this isn't even a medical book! However, the beauty of it all is that you are now *less* of a stranger to yourself—what you are made of, what makes you live and what brings you down! Since we've covered only your 'constructional aspects', it's time to get acquainted with the mechanisms that *actually* make you tick!

Heart Action. The action of the heart consists of successive cycles of contraction (systole), when blood is pumped out of the heart, and relaxation (diastole) of the muscular walls of the **auricles** (atria) and the **ventricles**. During the period of relaxation, blood flows from the veins into the two auricles, gradually distending them. At the end of this period, when the auricles are completely dilated, their muscular walls contract and force almost the entire contents through the designed openings into the ventricles. This action is sudden and occurs almost simultaneously in both auricles. The mass of the blood in the veins makes any backflow virtually impossible. The force of blood flowing into the ventricles is not powerful enough to open the semilunar valves, but it distends the ventricles, which are still in a condition of relaxation. The tricuspid and mitral valves open with the blood current and close readily at the beginning of the ventricular contraction which immediately follows the auricular contraction. The ventricular systole is slower but far more forceful; each spasm totally empties the ventricular chambers. As this contraction takes place, the heart is thrown forward and upward with a slight rotary motion. This impulse, called the **apex beat**, can be detected between the fifth and sixth ribs. The heart is entirely at rest for a short time after the ventricular systole occurs. The entire cycle can be divided into three periods: in the first period, the auricles contract, in the second, the ventricles contract and in the third, both auricles and ventricles remain at rest. With every beat, the heart emits two sounds which are followed by a pause. The first sound, coinciding with the ventricular systole, is dull and protracted while the second, made by the sudden closure of the semilunar valves, is shorter and much sharper. Diseases and other problems of the heart valves may change these sounds, while many other factors, including exercise, cause wide variations in the heartbeat—even in healthy people.

Pulse. At the moment of ventricular contraction, the blood entering into the arteries causes the artery walls to stretch. During diastole, the distended arteries return to their normal diameter—in part because of the elasticity of the connective tissue and in part because of the contraction of muscles in the arterial walls. This 'return to normal' is important in maintaining a continuous blood flow through the capillaries during the period that the heart is at rest. The expansion and contraction of the arterial walls, which can be felt in all of the arteries near the surface of the skin, is called the pulse. Your next query will probably be "does the heart rate change as we age?" The answer is an emphatic "yes". Everyone's pulse rate (average heart rate per minute) changes as we age. The adjacent table shows the trends.

Age	Pulse, beats per minute
Newborn	130
3 months	140
6 month	130
1 year	120
2 years	115
3 years	100
4 years	100
6 years	100
8 years	90
12 years	85
Adult	60 - 100

Sources of Heartbeat. Haven't you ever wondered what exactly it was that made your heart beat? How does it do so automatically, hour after hour, day after day—right from the time it started ticking in the mother's womb, till it finally stops at death? The answer lies in a very special group of cells that have the ability to generate electrical activity on their own. These cells separate charged particles while spontaneously leaking certain charged particles into other cells. This action results in electrical impulses being produced in the *pacemaker cells* that are spread over the heart causing it to contract. These cells carry out this action more than once per second to be able to produce a normal heart rate of 72 beats per minute. The rate and strength of heartbeat are controlled by nerves through a series of reflexes that speed them up or lower them. The contraction impulse, however, is not dependent on external nervous stimuli but arises in the heart muscle itself. Each phase of the cardiac cycle is associated with the production of an electrical potential that is possible to record (commonly known as an electrocardiogram, which is discussed in Chapter 10).

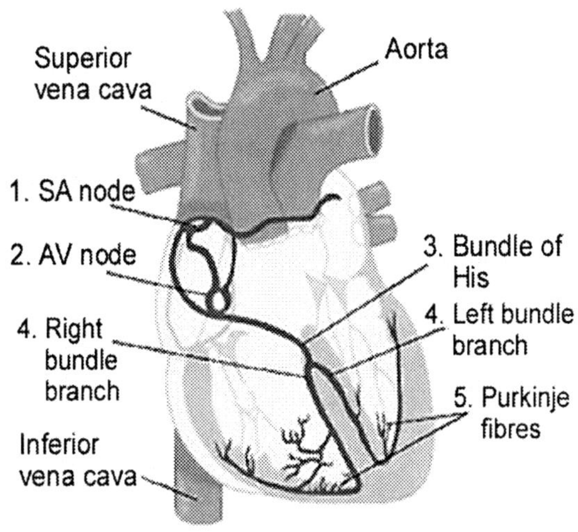

Figure 6 - Heartbeat Pacemaker Details

Let us follow through the heart beat signal process in detail using Figure 6. The heart's natural pacemaker is the **sinoatrial node (SA node)**[1] which is located in the RA. The heart also contains specialized fibres that conduct the electrical impulse from the pacemaker (SA node) to the rest of the heart. On leaving the SA node, the electrical impulse takes 0.04 seconds to travel to the right and left atria, causing them to contract together. At this point, a natural delay occurs that allows the atria to contract and the ventricles to fill up with blood. The electrical impulse has by now traveled to the **Atrioventricular Node (AV node)** [2] from where it goes to the **Bundle of His** [3]. From here the impulse then divides into the **Right and Left Bundle Branches**[4], where it rapidly spreads using **Purkinje Fibres** [5] to the muscles of the RV and the LV, causing both ventricles to contract at the same time.

Though any of the electrical tissues in the heart has the ability to be a pacemaker, it is the SA node, however, that generates the fastest electric impulse of all the tissues and so it is normally in control. If the SA node should fail, the other parts of the electrical system can and will assume control, although the rate of the electric impulse in this case will usually be slower. Electrical system abnormalities can range from minor premature beats (skipped beats) that do not require treatment, to slow or irregular beats that may need to be boosted by an artificial pacemaker.

Although the pacemaker cells create the electrical impulse that causes the heart to beat, other nerves can change both the rate at which the pacemaker cells fire as well as the strength with which the heart contracts. These nerves are all part of the **autonomic nervous system** which is further made up of the **sympathetic nervous system** and the **parasympathetic nervous system**. The *sympathetic nerves* are responsible for increasing both the heart rate and the force of contraction while the *parasympathetic nerves* do exactly the opposite.

The foregoing may have appeared difficult to assimilate and made you wonder why your insides function in such a seemingly complicated manner. Our intention is to convey the basic information needed for you to truly appreciate what a wonderful piece of machinery the human body actually is. The complexities of its various functions may puzzle you initially, but as we go further on and revisit these topics in later chapters, you *will* begin to view yourself in a new light and, hopefully, come to the realization that you really do need to look after yourself better.

But Does the Heart Also Hop, Skip and Jump?

Cardiac Arrhythmia. Arrhythmias are irregular heartbeats or disrupted heart rhythm or any change in the normal beating of the heart, which also includes an uneven pulse and a rapid, fluttery pulse. These are the most common of all heart conditions and **are actually the villains of the piece in over one million or so sudden heart deaths that occur the world over every year.** Electrophysiologists (clinicians and researchers involved in cardiac electrical function) have gathered conclusive proof that arrhythmias are the deadliest of all heart conditions, being directly responsible for more than two-thirds of all deaths linked to heart disease. In other words, these disturbances of heart rate and heart rhythm are caused when the heart tissue is unable to conduct the electrical impulse that regulates beating. Abnormalities of cardiac rhythm and conduction can be lethal. They are dangerous to the extent that they reduce cardiac output so that perfusion to the brain or myocardium is impaired. As you have seen earlier on, the sinus node is the heart's pacemaker, though this job can also be performed by any of the other electrical tissues. When the sinus node malfunctions, it is known as a disorder of impulse formation or automaticity. When the specialized fibres that conduct the electrical impulse to the rest of the heart malfunction, then a condition of abnormality of impulse conduction exists. Two other debilitating conditions are **re-entry** (a condition where there is an induced interruption in the circuit) and **triggered activity** (a condition where abnormal electrical activity persists after re-polarisation). In simple terms, if the heart is beating *incorrectly* it is, for all purposes, unduly straining itself *and also* disturbing the flow of blood to its own detriment. Cardiac rhythm disturbances are classified according to where they occur in the heart and their effect on heart rhythm. **Atrial arrhythmias** arise in the atria and **ventricular arrhythmias**—***generally acknowledged as the far more dangerous condition***—takes place in the ventricles. Arrhythmias can cause sudden death and are not to be taken lightly. This will be discussed in detail in Chapter 10.

Angina. Medically termed **angina pectoris**, this isn't a disease in itself but a symptom of disease. Often linked to coronary artery disease, this is a signal from the heart that it isn't receiving sufficient oxygen to work efficiently. From the physical standpoint, it is a sensation of pain or squeezing that radiates from the centre of the chest and extends either down into the left arm or upwards into the shoulders or the jaw. Angina lasts for a short duration and subsides completely without any residual discomfort. If the attack is caused by exertion and the individual promptly rests, the duration of the sensations would be less than three minutes. If the attack follows a heavy meal or is brought on by a bout of anger, the duration is normally between 15 to 20 minutes. If the angina worsens (more frequent pain episodes with less exertion), a condition known as **unstable angina** will then exist. This is caused when the blockage or obstruction gets worse. A heart attack occurs when a coronary artery is completely blocked and no blood or oxygen is reaching the heart muscle served by that artery. This will also bring about chest pain as well as cause the demise of the heart muscle served by that artery. Anything above 30 minutes is not only unusual but also indicates the development of unstable angina or a myocardial infarction. Angina in itself is reversible and does not cause death of the heart muscle cells.

Blood Pressure. This is the pressure of circulating blood against the walls of the arteries. Blood pressure is an important diagnostic index, especially of circulatory function. Because the heart can pump into the large arteries a greater volume of blood than that which can be absorbed by the tiny arterioles and capillaries, a resulting back pressure is exerted against the arteries. Any condition that dilates or contracts the blood vessels or affects their elasticity or any disease of the heart that interferes with its pumping power, affects the blood pressure. In a healthy animal, the blood pressure normal for its species is maintained within a certain average range with great constancy. Controlled by both cerebrospinal and sympathetic nerve centres, the complex nervous system mechanisms that balance and coordinate the activity of the heart and arterial muscles permit great local variation in the rate of blood flow without disturbing the general blood pressure. In 1996, scientists discovered that hemoglobin, the iron-protein compound that gives blood its red colour, also plays a role in regulating local variation in blood pressure. Hemoglobin was found to carry nitric oxide, a gas that relaxes the blood vessel walls, thus increasing blood flow. Hemoglobin controls the expansion and contraction of blood vessels, ***and thus blood pressure***, by regulating the amount of nitric oxide to which the vessels are exposed.

Blood pressure is measured at two points, namely (a) the high point, called **systole**, at which the heart contracts to empty its blood into the circulation and (b) the low point, called **diastole**, at which the heart relaxes to fill with blood returned by the circulation. Pressure is measured in millimeters (mm) of mercury by an instrument called a **sphygmomanometer**, consisting of an inflatable rubber cuff connected to a pressure-detecting device with a dial. The cuff is wrapped around the upper arm and inflated by squeezing a rubber bulb connected to it by a tube. Meanwhile, pulsations are listened to via a stethoscope applied to an artery in the lower arm. As the cuff expands, it gradually compresses the artery. The point at which the cuff stops the circulation and at which no pulsations can be heard is read as the **systolic pressure**. As the cuff is slowly deflated, a spurting sound can be heard when the heart contraction forces blood through the compressed artery. The cuff is then allowed gradually to deflate further until the blood is flowing smoothly again and no further spurting sound is heard. A reading at this point shows the **diastolic pressure** that occurs during relaxation of the heart. During a single cardiac cycle, or heartbeat, the blood pressure reaches maximum during systole and minimum during diastole. Usually, both measurements are given as a ratio expression of maximum over minimum, for example, 140/80. When a single figure is given, it is usually the higher, or systolic, pressure. In healthy persons, blood pressure increases from about 80/45 in infants, to about 120/80 at age 30, to about 140/85 at age 40 and over. This increase occurs when the arteries lose the elasticity that, in younger people, absorbs the force of heart contractions. Blood pressure is influenced by a wide range of factors and varies among individuals as well as in the same individual at different times.

One suffers from **hypertension** when the blood pressure becomes abnormally high. This happens when either arteriosclerosis or atherosclerosis is present. In both cases normal blood flow and circulation are hampered resulting in elevated blood pressure. High blood pressure can result in sweating, blurred vision, blinding headaches, rapid pulse and shortness of breath. This disease is brought on through cigarette smoking, stress, obesity, excessive use of stimulants, drug abuse and high sodium intake. Abnormally low blood pressure, or **hypotension**, is a result of infectious and wasting diseases, hemorrhage and collapse. Systolic pressure lower than 80 is usually associated with shock.

CHAPTER 4

FREE RADICAL ACTIVITY - THE BANE OF THE MODERN WORLD

What's So Free About Being a Radical?

Since, so far, we've trampled a bit on the ***medical*** and side-swiped the ***chemical***, we might as well go whole hog and get ***technical*** as well! Unfortunately, technicalities brook no shortcuts, and for you to better understand and appreciate what these miracle therapies really accomplish, it will be appropriate to first learn about *what* actually causes a disease or dysfunction. If you recall, all your "parts" are the sum of a whole and are programmed to work both in unison and in support of each other. However, since the heart plays the stellar role in this entire concert, it would naturally follow that any impediments in its smooth function would have an adverse effect on the roles of the 'cast and crew'. Doctors may tell you that no one knows what causes heart disease or they may just give details of the various risk factors, but most of them still won't admit that they are also as much in the dark regarding the specific understanding of how heart disease—or any disease for that matter—actually develops!

What is a proven fact is that heart disease starts young. When the United States Army ordered autopsies on soldiers killed during the Korean War, doctors were horrified to discover that 75 percent of these otherwise healthy young men—with an average age of only 22—had gross evidence of coronary heart disease. More recently, autopsies on American soldiers killed in the Vietnam War showed an even higher percentage of heart disease. In India, though the exact figures are not known, it is an accepted fact that the process of modernization has resulted in higher incidences of heart disease, specifically in infants and the very young. This is an extremely disturbing trend and, with more and more of our rural folk being exposed to the vagaries of urbanization, it is anticipated that these figures will further rise alarmingly because, firstly, the costs of specialized treatment are far beyond what the poor can afford and, secondly, as a direct result many cases are going to perforce remain unreported.

As we learned in chapter 3, the arteries that carry blood from the heart are constructed in three layers, the middle one being the muscle layer. This muscle helps control the amount of blood that moves along as chemical messages cause the muscle to expand or contract, thereby increasing or decreasing the amount of the flow. It is within this muscle layer that heart disease starts becoming serious. However, before *that* happens damage occurs to the inner layer—that part of the artery that is in immediate contact with the flowing blood. The prime cause of the damage to these two layers is the action of something called a **free radical.** Since this peculiar interloper exerts a profound effect on your ultimate health, we're going to 'zap you out' with the explanation. Free radicals play an important role in your body, both in health *and* disease. They have been known to latently influence many of the human disease processes—cancer, cardiovascular diseases, AIDS, chemical hyper-sensitivities, rheumatoid arthritis, allergies, microbial infections and ageing. On the flip side, as has been scientifically substantiated, free radicals are also ***vital to your health*** and are important instigators in the body's metabolic processes.

If the Radical Is Free - Why the Third Degree?

The smallest unit of an element is the atom. To understand better, the atom consists of a central ball, called the **nucleus**, that is encircled by an outer ring. Within this ring there are layers of one or more smaller balls, known as electrons, that orbit the nucleus in pairs. *Free Radicals*, also known as **reactive oxidant species (ROS),** are a chemical species capable of independent existence which have one or more ***unpaired*** electrons. Therefore, the free radical, which is so small that it cannot be seen even under a microscope, may be correctly defined as *a chemical species having an odd number of electrons and a single unpaired electron in the outer orbit.* Most chemical species have their electron orbits filled with paired electrons which spin in opposite directions, each balancing the other's magnetic moment. If a pair of electrons gets separated, an unbalanced situation is created. It is not the imbalance of the magnetic forces, which are too small anyway, that causes the disturbance, but *the occupation of the orbit by the unpaired electron.* Any non-radical becomes a *free radical* by either losing or gaining an electron. The single electron that occupies a whole orbit attracts a new electron partner and is easily excitable. Molecules that contain these uncoupled electrons are very reactive. A free radical persists in time only for a micro-fraction of a second but, within that period, it can cause a whole chain reaction by hitting some other atom or absorbing an electron and becoming neutralized by that action.

When an electron is ejected, a radical **cation** (positively charged) **atom** is formed and, vice versa, when an electron is trapped, a radical **anion** (negatively charged) **atom** is the result. Once free radicals are initiated, they tend to propagate and get involved in chain reactions with other less reactive species. As these chain reactions go along, they might—or might not—do some serious damage. It is possible for two radicals to collide and neutralize each other. But a single free radical could also hit a bit of rancid fat inside of you and cause millions of free radicals to be instantly created. The toxicity of a single free radical species may also be amplified in subsequent reactions. Free radicals are terminated by nutrient antioxidants, by enzymatic mechanisms or by recombination with each other. The life of a free radical, therefore, has three stages: *initiation, propagation and termination.*

To further simplify, a free radical is an **oxygen molecule** with an odd number of electrons in the outer orbital ring of one of its atoms. As already explained, molecules (and atoms) contain an even number of paired electrons. Why free radicals differ from all other molecules, ions and molecular complexes is *the single fact of the unpaired electron in their structure*—a distinction that may seem trivial, however, it has an enormous significance. If one of the electrons in a pair becomes separated, an imbalance is created which makes the resulting molecule (or atom) promiscuously unstable, violently reactive and very destructive—ready to aggressively attack any nearby substance and setting off further free radical reactions with explosive, cell-destroying power. What the free radical actually does is combine with and react chemically with another molecule that was *never meant to be interfered with!* Just as, on the outside, oxygen produces rust on metallic surfaces, so too do unbalanced oxygen molecules produce 'rusting' inside our bodies.

Free radicals are continuously generated as a result of many essential chemical reactions that occur naturally in our bodies. These reactions are necessary for life as they also are a part of the ongoing metabolic processes which are a by product of the normal use of oxygen— burning fuel (food) and generating energy—and are produced in large numbers in the **mitochondria**, the cell's complex 'oxygen reactor' or power plant. At the same time that oxygen and food are processed in the mitochondria, to generate the cell's energy requirements, a flux of free radicals is released. These interlopers head straight for the nearest target, the fatty acids, converting them into dangerous peroxidized fats which, in turn, produce even more free radicals! If they remain uncontrolled, free radicals can wreak havoc by breaking down delicate cell walls and damaging cells, by damaging important protein enzymes and by ravaging the sensitive structures of the mitochondria so that it ceases to be an efficient energy producer. Free radicals have a life span of just micro-seconds and their concentration in any site, at any moment, is minuscule—but they will attack anything in their vicinity with amazing speed.

The foregoing may initially look and sound like so much baloney, but if you go through it once again, it ***does*** make sense! So, are you interested in free radical illustrations? Certainly! The chlorine in your drinking water is full of them and you're gulping down mouthfuls every time you quench your thirst. Are you a habitual smoker? Then you're the original free radical factory! The very air that you breathe is full of them—especially the smog created by industrial units, motor vehicles and burning garbage. One of the worst sources is from rancid fats or heated oils. Every time you eat deep-fried foods, you instigate uncontrollable free radical activity. If you could but glance inside yourself, you'd be surprised to notice that many natural actions of the body also produce free radicals. Immune system cells, for example, handle bacteria by deliberately producing free radicals that immediately attack and kill the intruders. *Hence, though there are many important and beneficial aspects of free radical creation within the body, yet the body also tends to create free radicals* ***far in excess of what it actually needs for proper functioning.***

Remember the **trace metals**? Well, here's the connection. The greatest sources of free radical production are the tiny particles of metals—the size of a few atoms—that somehow get ingested. These include stuff like toxic lead and cadmium or abnormal copper and iron. You might like to flip back to Chapter 2 where we've already mentioned these and their effects. Thereafter it won't be hard to put two and two together! These elements don't get eliminated through the normal cleansing processes and, in a special form called ions, act as catalysts to the production of free radicals. You must be wondering why so much stress is being laid on free radicals. The main reason is that the understanding of *and* the findings on these are fairly recent, and it is becoming more and more evident that *the free radical is the most likely cause of all* *cancer and heart disease—and probably* ***every*** *other form of disease not caused by a bacteria.*

The Modus Operandi – Is It Also Radical?

There are six different activities of a free radical—all of them bad news for you! We'll tackle them one by one so you get a greater insight on how things deteriorate:

1. The nastier aspect of free radical activity is that just one of these chaps can, single-handedly, cause millions of his brethren to be created. What happens is that a free radical hits some fat, a susceptible material, inside the body and causes the fat to turn rancid besides triggering the creation of an untold number of other free radicals that go off in all directions at once. You might have some toxic metal inside you, such as lead. Don' be fooled into thinking that lead can *only* get inside you if it is physically consumed. Lead can get inside when auto exhaust fumes are inhaled, and, horror of horrors, the stuff remains inside you for years and years. When a free radical hits this toxic lead, the lead itself is not subjected to any changes but millions of other free radicals are immediately created. Thus, of all the things that a free radical can hit, hitting some tiny amount of toxic metal inside would cause the most harm.

2. Free radicals can cause two different particles to stick together—*things that ought* ***not*** *to be stuck to one another!* For example, you have the DNA, or the blueprint, for every cell. This, essentially, should be a separate particle inside the cell but a roving free radical can cause the DNA molecule to get stuck to some protein substance which in turn completely flummoxes DNA action. Though a lot more free radicals are ***not*** created when this happens, you can bet your life that the DNA molecule will cease to function as it originally did.

3. Cells depend a great deal upon the membrane, or wall, that keeps the outside stuff outside and allows the stuff *inside* the cell to remain ***separate*** from the stuff *outside*! This membrane is largely made of cholesterol but it has thousands of special parts which allow the outside stuff to go inside when needed ***or*** allow the inside stuff to go outside when needed! These are all very complex arrangements which allow the cell to take in food and

water as needed and to push out toxins and waste products that are not needed. A free radical can damage the integrity of the cell membrane or, in other words, *damage the membrane's ability to either pull the good stuff in or kick the bad stuff out!* Whenever the cells get threatened, their natural tendency is to manufacture more cholesterol to make the membranes thicker—a sort of 'added protection'. Some of this 'added' or extra cholesterol is bound to leak out, causing a resulting increase in the amount of cholesterol in the blood. Could *this* be the ***cause*** of heart disease? There is much of modern medicine that confuses *effect* with *cause.* Don't you dare fall into the same trap! When a free radical attacks a cell, the cell counters by creating more cholesterol. Doctors see this extra cholesterol and decide that cholesterol is the source of the problem; **IT IS NOT**! This extra cholesterol is nothing more than a *symptom* of some *real cause—**free radicals attacking the cell!***

4. Inside your body there's a fancily-named juice that finishes the digestion process if *that* didn't get thoroughly done in the stomach. Free radicals can cause these special digestive juices to invade some cell and literally start consuming the cell from the inside. The fancy stuff is called **Lysosome** which is a minute body existing in many types of cells. It contains various **hydrolytic enzymes** and is normally involved in the process of localized intracellular digestion. Injury to a lysosome is followed by the release, into the cell, of the enzymes which may damage the cell and give rise to wasting and other pathological aspects of certain diseases, as in muscular dystrophy.

5. As you grow older, "age spots" often tend to appear on your skin. These are sometimes called *liver spots.* This is nothing more than a pigment or dye created by a free radical penetrating into a cell in the skin. The pigment is created inside the cell and the cell takes on some colour. Though the damage to the cell's function is relatively minor, there still *is* damage. As more and more cells get affected and the colour deepens, the damage gets greater. It is not only the cells of the skin but any cell, anywhere in the body, that can get coloured and damaged in this way.

6. A free radical inside a cell can damage the blueprint, or DNA, by which that cell creates new cells. This is the main cause of cancer and the beginning of heart disease.

Are Things Really That Bad?

Overwhelmingly detrimental as it may appear, actually not all the news about free radicals is bad news. Surprised? Read on. Free radical reactions are unavoidable because of their involvement in metabolism in biological systems. Normal metabolism contributes to the production of those free radicals used by the body to produce beneficial effects. Emotional stress may also contribute to free radical production. Oxygen-containing free radicals participate in the metabolic processes essential for life. The white blood cells, involved in the body's defence system, produce and mobilize free radicals of oxygen to destroy bacteria, fungi, viruses and other foreign matter which they ingest. Free radicals govern the production of **prostaglandins** which are vital for the protective and physiological functions of the body. The liver, too, uses free radicals in the process of detoxification. The total number and types of free radicals are unknown, but an estimated 10,000 hits from free radicals bombard each cell every day. With the body containing about 100 trillion cells—*that's* some free radical activity! Obviously, if *all* free radicals were deadly, a person would cease to exist and, therefore, it is important to remember ***that antioxidants be taken in moderation if and when they are used.***

Some necessary 'enlightenment' on antioxidants: They may be defined as any substance that, when present at low concentrations in comparison with the concentrations of an **oxidizable substrate**, significantly prevents or delays any oxidation of that substrate. What the word means is "against oxygen" and it has been so coined because the most common form of dangerous free radicals are those created from oxygen. We *must* have oxygen to live and we're breathing it in all the time, but it's a fact that an atom of oxygen—usually so necessary and vital to our lives—can also actually become a source of disease and death. Since oxygen is the most common form of a free radical, the method of neutralizing all free radicals has taken on the name "anti-oxidant"; however, these substances only oppose the free radical form of oxygen. In recent times antioxidants have elicited great enthusiasm and interest in the food industry because they prevent rancidity. Oxidizable substrates include almost anything found in foods and living tissues including proteins, lipids, carbohydrates and DNA. They are of wider interest to society because of the important role that they play in protecting the body against damage by free radicals. Thus the key issues become target selection and the source of oxidative damage. When reactive oxygen species (ROS) are generated in living systems, a wide variety of antioxidants come into play. The relative importance and efficacy of these in the body depends on which ROS is involved, how it is generated, where it is generated and which target of damage is selected. What needs to be remembered regarding antioxidants is that they may offer protection to the body in one system but fail to do so in other systems. In certain circumstances, an antioxidant may not only ***not*** have a protective action but may even aggravate the situation. Antioxidants that are protective in foods may not necessarily be protective in the body. That is why you are cautioned to use these in moderation.

Good Radical? Bad Radical? Surely the Next is Ugly!

You couldn't be more right! We had mentioned that a free radical inside a cell can damage the blueprint (DNA) by which that cell creates new cells—which is the main cause of cancer as well as the beginning of heart disease—so we're going to delve a bit more into this phenomenon. The DNA inside a cell is the plan that tells the cell how to divide into new cells and when to do it. It is an extremely complex molecule made up of millions of parts. In fact if one were to string all these parts in a straight line, the DNA of just *one* cell would stretch out about six feet! Since we possess an amazing 100 trillion cells in the body, each having that six-foot long DNA strand, you'd go nuts trying to figure out the mind-boggling mileage of DNA that a single person possesses! Since a free radical is smaller than microscopic—the size of a single atom in many cases—the one way that it causes damage is that when it enters the cell in the quest of a mate it attaches itself to some other atomic particle instead. The cell, of course, has all sorts of things inside; many of these items are combinations of atoms with a few spare electrons circling on the outside rings. Take, for example, the hydrogen atom as a free radical. All it takes is for this free radical to attach itself to some ***other*** particle inside the cell and this would be sufficient to cause that particular particle to malfunction and kill the cell.

That wouldn't be too bad in itself, except that, more often than not, something else occurs. *It may already have happened to you!* Every cell is implanted with this DNA blueprint according to which similar cells are reproduced. A free radical can alter the blueprint by just adding an electron ***where that electron doesn't belong***. Since that DNA has millions of atoms, any one of these could be responsible for the blueprint producing a cancer, instead of a new healthy cell, the next time the cell divides! Whenever the blueprint is subjected to an alteration—for whatever reason—the cell invariably initiates a 'wrong' production. It could reproduce into a defective cell that dies quickly ***or*** it could reproduce into a defective cell that will continue to grow in some form of cancer. And ***THAT*** is how cancers and other strange aberrations are created in the body. Therefore, any alteration in the DNA blueprint of any cell will have serious detrimental consequences.

There's More! There's Much More!

When, somehow, there exists the situation of free radicals gallivanting inside, invariably these gain access to the blood where millions more get manufactured. They can cause damage to the inner lining of an artery, thereby creating a sort of malleable passage through which more free radicals easily penetrate into the cells in the muscle layer of the artery. The cells of the artery muscle are particularly susceptible to free radical damage. If the damage happens to be in the DNA, the new muscle cells will start being created far too rapidly and metamorphose into a **tumor**—which is an abnormal growth of cells that is *not* poisonous or is non-malignant. These are termed as benign tumors because they are not harmful in the long run. That may be well and good, but harmful or not, ***they shouldn't be there in the first place!*** As the muscle cells grow rapidly and become tumors, the growth has nowhere to go but to intrude and poke out into the space where the blood flows through the artery. In other words, there is now a bulge from the middle layer of the artery, pushing inwards through the inner lining. If this bulge kept on growing, it would eventually choke off the blood flow. But that doesn't happen because there is a natural limit to the size of these things. These growths need fresh blood to keep them alive and, since they don't develop their own blood network, the only supply available is the blood seepage through the mass of cells, which isn't enough to encourage unlimited growth.

As the tumor grows out into the blood flow area, breaking through more of the inner lining in the process, the body senses the danger and triggers one of its many defensive systems to activate and counter the threat. Immediately, a protective layer containing cholesterol starts getting deposited over the damaged area in the artery. Cholesterol is one of the most defensive substances that the body possesses and is so important that every single cell has the capacity to manufacture it when attacked *AND* to spread a protective coat around the damaged area. It is not exactly the bulge that causes a heart attack. If the bulge *were* to grow to a size that totally blocked the artery, then it *would* be responsible—but we know that the size is limited.

One of the main causes of heart attack is the *hardening* of the arteries. As our arteries get harder—the more useful term would be **less elastic**—they are less able to contract and expand to the body's needs for less or more blood. As they lose the ability to tighten and loosen, there comes an occasion when the body needs more blood; the artery cannot expand to accommodate the larger flow and you have a heart attack! The real cause of hardening of the arteries is *not* that the extra coatings of cholesterol attract extra calcium which may eventually accumulate at the site. The real reason is that the tumor reaches a size where the blood supply can no longer reach its central core which is ultimately choked closed—so the tumor dies. The process of death is such that the cell gradually loses its ability to rid itself of toxins—including calcium—and turns into a sort of solidified mass of calcium! As roundabout as that sounds, it is nonetheless an absolute fact!

Some folks tend to refer to heart disease as the malaise caused by the build up of cholesterol in the arteries. This is only partly true and tends to overlook the real villain of the piece—calcium! The arteries don't get hard until calcium starts to build up, which it actually does INSIDE the cells. Prior to this, even though a tumor may be in existence as a result of free radical damage, the arteries continue to remain flexible enough to permit unimpeded blood flow. The triggering device for heart disease is the build up of calcium ***inside* the individual cells within the tumor** causing that area to become rigid and lose flexibility. Cholesterol, on the other hand is a soft, buttery substance and, even if it happens to get collected in large amounts, it is capable of being washed away by the blood flow since it cannot harden. However, when it mixes with calcium, *it is the combination that hardens*, which is what happens to the centre portion of the dead growth within the artery—it becomes a rock of calcium glued together with cholesterol. As time marches on, the progressive hardening of the arteries inhibits expansion for increased blood flow and that part, where the dead tumor is now a rock-like formation, will naturally have a narrower passage. If a tiny chunk of clotted blood, normally en-route for ejection from the system and the body, happens to flow through this passage and hits the obstruction the artery can be completely plugged. That's when you're likely to have a stroke.

The Calcium-Cholesterol Combine

Since you are quite clear regarding 'how' and 'what' happens, you might as well recognise and be familiar with the medical terms for the arterial disease processes. There are two main titles that are commonly used and which cover practically all the aspects that we have gone into some length to describe—they are also fairly easy to differentiate! With the passage of time, calcium deposits build up and, in association with cholesterol, form calcified and atherosclerotic plaque which starts lining the arterial vessels. When **calcium** predominates, the process is called **ARTERIO**SCLEROSIS or *hardening of the arteries.* When **cholesterol** predominates it is called **ATHERO**SCLEROSIS. The exact content of the plaques is determined by an individual's diet, antioxidant intake and the duration of the process. Regardless of where on the arterio- / athero-sclerosis continuum any particular individual falls, the result is the same—**less and less fresh oxygen is delivered to the tissues of the body.** The development process is as follows:

> **Arteriosclerosis.** Hardening of the arteries or arteriosclerosis (as distinct from *atherosclerosis*) is apparently an inevitable change due to ageing. The walls of the blood vessels become stiffer—as do all connective tissues of the body—with the passage of time. This is caused, firstly, by the cross-linkage of **collagen** (the protein which makes up the connective tissue of artery walls) and, secondly, by the slow and steady diffuse deposition of calcium in the walls of the body's arteries. This calcium then builds up and becomes many times more concentrated than it was during childhood. We are not talking about plaque formation but rather a diffuse deposition of calcium in the walls of the arteries, finally resulting in an arterial system that may become as stiff as a lead pipe! Calcium content is what both *arteriosclerosis* and *atherosclerosis* have in common, but in the former it occurs diffusely while in the latter it occurs in concentrations called plaque. The most 'popular' result of calcium deposit is a ***heart attack!*** This is also known as a **myocardial infarction**. In most heart attacks both athero- and arterio- sclerosis are present. **Athero** provides the plaque which narrows the artery while **arterio** jumps in to stiffen the artery enough so that it cannot expand to compensate for the blockage caused by the plaque. Thus occurs what one might term a dual assault on a single target!

It would not be incorrect to consider ageing as the progressive dysfunction of calcium metabolism. As the body ages, more and more calcium is concentrated in the interior of the cells. Dying of old age, in reality, is dying of calcium poisoning! Recent evidence on the metabolism of **homocysteine** emphasizes the importance and implication of this amino acid in speeding up the ageing process if it occurs in higher concentration levels. While **arteriosclerosis** by itself ***does not*** cause a heart attack or stroke, the condition enhances the effect of **atherosclerosis** to produce these results. Potentiation, or enhancing, occurs by virtue of the fact that when plaque develops in the presence of **arteriosclerosis**, the stiffened and inflexible blood vessel wall does not have the ability to compensate by expanding with the increased blood pressure of each heartbeat. Thus the flow of blood is less than otherwise due to absence of flow around the plaque.

Atherosclerosis. The cells' ability to move metal ions through the system and eliminate them, when they are in excess, becomes progressively impaired in the presence of ageing and disease. This is especially true for calcium which has vital functions in the human body. Without it, teeth and bones could not exist. Nevertheless, as the body ages, oxidation of lipids damages the walls of the artery which are repaired with fatty substances leaving a scar. Calcium and oxidized cholesterol are incorporated into the resulting scar tissue. The resulting lesion is called an **atherosclerotic plaque** and the disease process is known as **atherosclerosis**. With this disease, as the years pass, calcium deposits build up and calcified plaques form which line the walls of the arterial vessels. This plaque is composed of various lipids (so-called foam cells), scar tissue and overgrown smooth muscle cells from the artery wall. In many people this process begins in early childhood. When this process affects the blood vessels supplying the brain, strokes or transient ischemic attacks may occur. A stroke involves at least temporary non-function of certain brain cells—which may become permanent if oxygen doesn't arrive in time. Transient ischemic attacks are dizzy spells caused by poor circulation to the brain. Conventional medicine has **nothing** to offer that can reverse the ravages of atherosclerosis.

No doubt that there's been plenty of ebb and flow in this text so far and we've also had to dwell a bit longer on certain aspects—perhaps, from your point of view, we've also gone into lengthy descriptive spiels! However, it was all a necessary evil! What we *are* quite sure about at this point is that you are totally familiar with the entire picture and, therefore would now be in a better position to appreciate exactly how the new age therapies will benefit mankind in the long run.

Some 'Pondering-Purpose' Pointers!

In the presence of ageing and disease, the cells' ability to move metal ions through the system and eliminate them when they are in excess, becomes progressively impaired. This is especially true of calcium, which has vital functions in the human body. Without calcium, teeth and bones could not exist. Nevertheless, as the body ages, **lipid peroxidation** damages the walls of the arterial tree which is repaired leaving a scar. Calcium and oxidized cholesterol are then incorporated into the resulting scar tissue. There are several known, *and* easily avoided, risk factors at work in the creation of arteriosclerosis. Lipid peroxidation begins the inflammatory process in the wall of the artery and is facilitated by the presence of 1) **polyunsaturated fatty acids** that are present in many of the 'junk foods', 2) oxidized cholesterol from pasteurized milk and other animal foods cooked in open air, 3) the relative absence of antioxidants such as vitamins A, C and E and high levels of *homocysteine*. Tobacco smoke drains the body's resources of antioxidants, particularly vitamin C, and further accelerates arteriosclerosis. If you know and apply these facts from an early age, there is no reason for arteriosclerosis to develop in your body. Hardening of arteries is not an inevitable disease of ageing, as you may have been led to believe—*it is a disease of bad habits!* To know and apply these facts, you have to be ***willing*** to think for yourself—something that seems to elude the normal individual, especially where certain important medical aspects are concerned, and ***that*** is the reason why arteriosclerosis will continue to kill people. Maybe the good do die young—but so do the *uninformed* and the *dogmatic*!

In years gone by, it was once a medical opinion that arteriosclerosis was a process that began only in middle or old age. ***The latest findings confirm that it begins in childhood in a large percentage of the populace.*** The severity of this life-long process is determined by genetics, level of exercise and dietary habits. By the age of 21, many individuals have already been struck by arterial disease! This is definitely a disease of modern civilization. Never before have people so young been afflicted, in fact, as recently as the year 1900, heart disease was extremely rare. Now we ***know*** that airborne industrial pollutants, herbicides, pesticides and preservatives in our food are some of the major culprits responsible for this outrageous "theft of health". The cholesterol content of plaques can be effectively handled by a change over to a no-fat, high-fibre diet. Plaques actually decrease in size and the cholesterol content can eventually disappear. Lipid peroxidation itself can be halted by a liberal intake of antioxidants such as **beta-carotene** (the precursor of vitamin A), mixed **tocopherols** (vitamin E) and vitamin C so that no further damage is caused to the arterial tree. The calcium content of the scar-plaques already present is another matter. Therefore, if you want to restore your health to a completely youthful condition, you face a real challenge with arteriosclerosis.

The list of problems that can be caused by artery disease is really impressive, but the extent of the maladies should not come as a surprise to anyone—considering the fact that a fresh and continuous supply of oxygenated blood is absolutely necessary for the proper functioning of any organ. Even diseases that are more complicated, in that they have causes other than decreased blood flow, are made worse by arteriosclerosis. Diabetes gets worse due to poor blood flow to the pancreas. Not only that but this poor blood supply is also the cause of the decreased output of digestive enzymes from the exocrine part of the pancreas resulting in incomplete digestion. In the kidneys, decreased blood flow results in the inappropriate release of **angiotensin** by these organs which results in the inducement of hypertension throughout the vascular tree. The joints, particularly the joints of the lower back, react with pain and inflammation to the decreasing blood flow and this, coupled with the degeneration of ligament tissue and disc disease, is responsible for the so-called *"low back syndrome"*.

Arterio- and athero-sclerosis also play a major role in the spread of arthritis throughout the body. The effect of this process on the heart is angina and eventually an infarction and death. Poor blood supply to the stomach and small intestines results in poor digestion. Poor blood supply to the colon causes slowing of the colon and can end up as colon disease. The effect on the extremities is cold hands and feet. Impotence is also caused by a decreased blood flow to the penis. Frigidity results when the pelvis gets insufficient blood. Cancer has been known to accelerate if there is a decreased blood flow to the affected tissues. When the blood flow to the immuno-competent cells in the bone marrow and spleen gets reduced, the immune system itself is weakened. The list of aches, pains, discomforts and diseases caused, or made worse by, arterio-/athero-sclerosis is never-ending.

Why the Obvious Stress on Blood and Blockages?

The last sentence above should be a pointer of sorts! The essence of human existence is the blood that flows in the system. It is responsible for both rejuvenation ***and*** enervation. In simple terms, 'rich' blood ensures rich physical standards while 'poor' blood means a lifetime of ill health. Most of us pay little or no heed to the physical condition and properties of our blood. We would probably be more alarmed if our physician predicted an outbreak of measles than if he informed us that the red blood cell count had decreased drastically! It is almost a case of "***what the eyes can't see, the heart won't grieve over***". It is hoped that through what you have read so far, the realization *has* dawned that absolutely everything regarding health is irrevocably connected to the overall condition of the blood. We also invariably place the heart on the highest level in the hierarchy—and probably not without reason. Isn't it the only organ that *never* rests? That may be true, but it also should be kept in mind that the heart's ***sole and only*** function is to pump blood through the entire body. If blood conduits somehow get blocked, then the pumping action suffers a reversal . If it is an inlet conduit, then the correct amount of fresh, oxygenated blood isn't reaching the heart resulting in poor oxygen supply, poor cell generation, weakened heart muscles and pumping dysfunctions. If the outlet conduits get blocked, then oxygen supply to the rest of the body is similarly adversely affected. The same may be applied to all other organs and systems within us. Of course, it is an extremely complex inter-dependency but there are certainly no doubts that the blood is gaining more and more importance in the overall scheme of things.

Therefore, laying a lot of stress on understanding the functions and properties of the blood, and the effects and reasons for blockages, will not only definitely benefit the viability of our plans for the maintenance and propagation of health, but it will also help to steer us in the right direction ***when we are in need of medical help and advice***. In a nutshell, if the purity of the blood is maintained in a pristine condition, the body's own defence and recycling mechanisms would probably work more efficiently without 'pollution'— in the form of any man-made, back-up stimulation and medications! Unfortunately, modernization has demanded—and extracted—its pound of flesh. Not a single living being can lay claim to being totally untouched by scientific advancement. The rot has established itself so deeply that we humans could possibly be termed as an "affectedly infected" species, susceptible to infections, dysfunctions and deformities from the very foetal stage. The end result of ignorance and nonchalance is that we have used and abused life's greatest creation—the human body—in so many vile ways that many of us have reached the end of a road ***that is yet to be charted and laid!***

Where Do We Go from Here?

This question brings this narrative to **that** point, which necessitated all those details in the first place, where we will now positively have to bury all pessimism and look only optimistically at the future. You will now get to know, and learn much more, about the miracle therapies that we introduced you to at the start. A word of caution: if you have skipped to this section first, then you need to go back over the previous pages if you are to make any sense of the intricacies that we will be describing—in addition to understanding fully what benefits are to be accrued from and the working extent of each therapy.

Some individuals may contend that they are in the prime of life and enjoy exceptionally good health so there is no real need for them to know or learn more at this moment in time. We hate being spoilsports, but the fact of the matter points to over 90 percent of us urbanized folks already having been successfully targeted by disease. We had mentioned earlier that "what the eyes can't see, the heart won't grieve over," and with all those **free radicals** already having made inroads and working away feverishly, it's only a matter of time before that unseen uppercut lands on an unsuspecting and unprepared jaw! Why must you wait for an illness to manifest itself and only then resort to treatment? It is far, far better to learn about the hidden dangers and take the appropriate steps to ensure that disease can never strike!

These pages are not only a 'wake up' call but are also intended to act as mentor and guide. Since every single one of us is currently under some sort of attack or the other, no matter how healthy one may outwardly appear, it is time to take stock and make the necessary preparations to prevent our physical selves from being overwhelmed. Individual chapters have been devoted to each therapy and these will be followed up by a chapter on how two or more of these therapies can be successfully used as a combination for enhanced healing aspects and longer lasting benefits. Prepare then, to enter into a truly magical realm. If you're ready—so are we! Welcome to the world of miracle treatments and a sensible approach to good, lasting health!

chapter 5

CHELATION THERAPY – THE AMBIDEXTROUS HEALER

Which Way to Turn?

RT was only 46 years old, but his physical condition was so bad, he may as well have been 86! Ten years of enduring diabetic hardships, contending with unstable angina and a positive diagnosis of **triple vessel disease (TVD)** had taken their toll. Blood tests also confirmed grossly abnormal fasting sugar and **triglyceride** levels. He was regularly in and out of hospitals and had to rely on a series of medications that, far from providing relief, only made him feel weaker and more ill. He developed cataracts and needed surgery to facilitate vision. Eventually a stage was reached when he became totally incapacitated, unable to walk or perform the simplest of tasks without help. His doctor could provide neither worthwhile medication nor any further relief and had given up hope. Fortunately, RT found out about chelation in time. He came to our clinic in a truly pathetic physical state and looked far older than he actually was. Chelation therapy was commenced right away and 18 sessions later he was able to resume normal mobility to a limited extent. After 35 treatment sessions, RT regained complete mobility and was able to return to work. Whereas he initially couldn't even move, he now commuted daily to and from work and easily climbed the four flights of stairs to his office without any problems or complaints—*or help!* The angina problem just vanished completely, clinical re-tests for TVD revealed that his arterial blockages had cleared up to a considerable extent and a post-therapy thallium stress test showed tremendous improvement. Though his glucose and triglyceride levels remained somewhat abnormal, they had certainly reverted to more acceptable levels. After 51 therapy sessions, he has completely resumed a normal lifestyle and comes in once every three months for a 'maintenance' session of chelation therapy.

Attaboy Chelation!

Chelation is a chemical reaction that results in a bond being formed between a metal ion and an organic (mostly carbon) molecule. The resulting complex—metal bound to molecule—is called a **chelate** and contains one or more rings of atoms in which the metal ion is so firmly grasped that it cannot escape. This allows the metal ion, in this 'captured' form, to be transported through the blood stream to the kidneys to be eliminated.

The chelation process goes on constantly in nature. It takes place in the bodies of living organisms, both plant and animal, as a natural function. Chelation is the means by which plants and animals are able to utilize inorganic minerals. For example, **chlorophyll**, the green matter of plants, is a chelate of **magnesium. Hemoglobin**, the oxygen-carrying pigment of human red blood cells, is a chelate of **iron**. The formation and function of enzymes, the protein substances which control most of the functions in the body, are all products of chelation. **Aspirin**, **citric acid** and **cortisone** are known to function, at least partially, as chelating agents.

The chelation process may, in fact, be responsible for the effectiveness of almost all the drugs used in medicine to overcome disease. Without its help, they would be unable to act and react with the body's physiology. Some of the most basic, and yet complex, chemical reactions found in nature—and in man—are encompassed by the chelation process. As a process in plant and animal biology, this reaction has been recognised for over half a century. It is a biochemical process which is fundamental in:

- Food digestion and assimilation.
- The formation of numerous enzymes.
- The functions of enzyme systems.
- The detoxifying of certain (toxic) chemicals.
- The synthesis of many hormones.
- The detoxifying of certain metals such as lead, arsenic and mercury.
- The movement of vitamins, minerals, hormones and other nutrients across membranes of body tissues as part of transport systems.
- The protein catalyst that controls most of the body's vital biochemical functions.

These very principles are applied in chelation therapy to treat the hardening of arteries and the diseases resulting from this effect. Taken from the Greek word **chele** meaning 'claw' (as the pincers of a crab or lobster), the term **chelation** refers to the way certain synthetic chemical and body proteins can bind metal molecules. The usual metal molecule possesses one, two, three or more positive electrical charges. The chelating agent has extra negative charges that can combine with the positive ones of the metal and hold it fast in the clutching grip of the 'claw' (chele). This gripping combination or *chelate* takes on entirely different properties than the metal alone ***or*** the chelating agent alone. In effect, an entirely different chelated compound is formed whose atoms are firmly held together.

The natural process of chelation is ongoing in the body. Iron, for instance, is a necessary metal present in your metabolic processes and its transportation in and out of your cells is handled by the chelation mechanism. This means that the dietary iron, consumed in food, is chelated within the tissues of the body. Although tightly held in hemoglobin, in the red blood cells and the body's oxygen carrier, iron may be transported through the blood stream and dropped where it is needed—in a tissue, organ, cell or other part of the body. The binding, of a metal such as iron, is very sensitive to changes in body temperature, acidity, concentration of metal, concentration of chelating substance, presence of other metals and the presence of other body chemicals. Such changes may bring about the release of iron and exchange it for another metal or for the binding of more or less iron molecules. In this way, the naturally present chelating substance, hemoglobin, can pick up iron and other metals from one location and, besides transporting them to another, is also readily able to release them when the local tissue factors change.

What, Exactly, Is Chelation Therapy?

It is a medical treatment that improves metabolic and circulatory functions by removing toxic metals (such as lead and cadmium) and abnormally located nutritional metallic ions (such as copper and iron) from the body. This is accomplished by administering a synthetic amino acid, called **ethylene-diamine-tetra-acetic acid (EDTA)**, through an intravenous infusion. **EDTA** is a man-made, protein-related amino acid molecule with unique and valuable therapeutic properties—one of which is its powerful attraction for loosely bound metals, especially those that speed ***free radical*** damage. EDTA pulls abnormal metal ions out of the body, thus reducing the production of ***free radicals*** and preventing the conditions under which 'scum' builds up on blood vessel walls. EDTA has evinced a great affinity for metallic minerals that have two positive electrical charges, like lead, with which it combines very well. Infusion with EDTA (chelation therapy) is the established method used world-wide for the removal of heavy metals—such as mercury, cadmium, nickel, copper, lead and others—from persons with toxic concentrations of these in their tissues.

But Aren't There 'Good' Metals Too?

That's right! In fact, one might vehemently opine that it is dangerous to have essential nutritional metals such as iron, copper and zinc removed from the body. As true as this may be—so long as they ***are*** where they ***are supposed to be***—essential metals are so tightly bound to the sites of normal activity that they are not easily removed. The metallic ions most apt to be removed by EDTA are those that are loose and free-floating and generally **ARE** where they are **NOT** supposed to be! Toxic metals, such as lead and cadmium, are truly double trouble for the body because they inhibit vital enzyme function (disrupting basic metabolic processes) and also increase *free radical production.* EDTA binds tightly to many abnormally placed metal ions. It forms a chemical bond that facilitates removal of these metals from sites where they may interfere with normal metabolism or enter into an undesirable chemical reaction. In the early days of its usage, EDTA action was erroneously thought of as the process of reaming calcified plaque out of the arteries and thereby helping to increase the blood flow. Although EDTA is definitely a calcium chelator, its affinity for calcium is far lower than it is for metals such as iron, copper, cadmium, lead etc. It will quickly drop calcium to pick up any one of these other metals.

How Did It All Begin?

EDTA was never developed to chelate metals from the human body but, incredibly, *to chelate calcium stains from textiles.* It was first produced in Germany in 1930 as a substitute for citric acid. Its development formed part of Germany's plans and efforts to become an independent producer and international supplier of citric acid. No one dreamed, at that time, that it would ever be of any medical use. Actually, it was some in-depth research, originated in Europe in the 1940s, which eventually led to EDTA's current biological applications. In those days, lead poisoning was quite common in certain industries, and it was discovered that EDTA could chelate lead out from those who had somehow infused lethal doses of this poison. Among the groups who were most frequently treated for lead toxicity were those who worked in battery manufacturing factories and also workers who spent more than reasonable amounts of time utilizing lead-based paints. Because of their close and lengthy association with a dangerous polluting agent and free radical producer, many of these people also suffered from arterial disease. Once the lead had been successfully chelated out of their systems, doctors were amazed to discover that many of their artery-based problems had dramatically improved as well. This led to EDTA's value being firmly established as a reverser of arterio-/athero-sclerosis.

However, more than just the arterio-/athero-sclerosis connection, chelation therapy has proved that it can increase blood flow throughout the body. It is also known to have improved liver function, improved the memory, reduced incidences of cancer, improved blood cholesterol ratios, improved vision, reduced blood pressure, lowered blood fats, relieved angina pains, forestalled heart attacks and strokes, relieved symptoms of senility, reduced leg cramps, relieved symptoms of arthritis, Parkinson's disease and multiple sclerosis, and healed ulcers that were caused by poor circulation. The versatility and efficacy of EDTA is certainly beyond question. The formidable array of improvements it provides, however, will not appear out of the ordinary at all when one considers how vital to health the arterial system truly is. Since it extends to every inch of the body, inclusive of every organ, the arterial system ***must*** exert multifarious effects on health. On the question of the exact extent of increased blood flow resulting from chelation, all that you really have to know—and accept—is that the increase is sufficient enough to relieve symptoms, promote healing ***and*** improve the quality of life. It takes only very slight vascular changes to significantly alter blood flow. Of the thousands of arteries in the body, most are so slender that their tubular dimensions are no larger than a human hair. Some capillaries are so narrow that single blood cells must fold themselves in half to be able to squeeze through! But enough of blood flow, let's see what effects EDTA has on enemy No. 1—***the free radicals***.

Since you're already familiar with free radicals you know exactly how they exert harmful influences. In a healthy body, free radical reactions are not only controlled but are 'allowed' to proceed in an orderly fashion—as dictated by the needs for energy production and for the detoxification of chemicals, germs and foreign substances. There are several enzymes that prevent the free radicals from running wild. When functioning properly, these enzymes work in concert with an elaborate system of natural free radical scavengers—***which include the antioxidant vitamins C and E***—that dampens free radical chemical reactions. Thus the desired biological effects are permitted to take place without any unwanted cellular or molecular damage. If such controls were not exercised, free radical production would run riot—something akin to a nuclear chain reaction where out-of-control free radicals are being generated at ever increasing rate. When this happens, they disrupt cell membranes, hinder the transfers across cell membranes, damage essential enzymatic proteins and also cause mutagenic damage to the genes and chromosomes. All these activities produce very sick and even ***malignant*** cells.

If the rate of free radical production proceeds unrestrained, the damage will be cumulative and progressive, eventually neutralizing the body's natural defences and leading to the million-fold, explosive increase of free radical concentrations. The resulting cell destruction, malignant mutation and damage to enzymes will cover the entire spectrum of circulatory, inflammatory, malignant and immunologic disorders that are largely the major cause of age-related illnesses. The link between free radical activity and malignant change was recognised decades ago and is currently the basis of much cancer research. The impact of the acceptance of the free radical theory—**free radical pathology**—could provide deep-rooted implications as did the ***germ theory*** of a century ago. Just as this theory provided a scientific basis for effective treatment of the infectious diseases, the major killers of the nineteenth century, similarly the ***free radical concept*** will help to formulate a solid platform from which effective prevention and treatment will ultimately emanate for the major killer diseases of the new millennium. As it is, there is a recent suggestion that the major chronic diseases of the modern world—cancer, arthritis, senility, atherosclerosis and related circulatory disorders, AIDS etc.—may actually be forms of radiation sickness resulting from continuous internal reactions, a sort of 'meltdown' of cellular power plants leading to accelerated ageing, ineffective defences and premature death.

But what, you may ask, has all this to do with chelation therapy? The obvious, instant reply is ***EDTA drastically reduces free radical production***! It is virtually impossible for free radical pathology to be catalystically accelerated by metallic ions in the presence of EDTA.

Unbound metallic catalysts must be present for an uncontrolled free radical proliferation to take place. EDTA binds ionic metal catalysts, making them chemically inert and removing them from the body. Therefore, chelation stops excessive and uncontrolled free radical production in its tracks, halts the development of free radical-inspired diseases and allows the body to repair what damage it may have *already* suffered.

What Exactly Can Chelation Correct?

Without going into any chemical/biological processes, chelation has been successfully applied in cases of heavy metal poisoning and atherosclerosis. In addition it is extremely efficient in correcting:

- Alzheimer's disease
- Aneurysm, cerebral ischemia, stroke.
- Calcific tendonitis, calcific bursitis
- Cancer
- Cataracts, glaucoma
- Cerebral arteriosclerosis
- Diabetes, diabetic ulcers, diabetic gangrene, diabetic retinopathy
- Emphysema
- Gouty arthritis
- Heart arrhythmias, hypercalcemia.
- Heart valve calcification.
- Hypertension
- Kidney stones and gallbladder stones
- Leg ulcers, snake bite (poisonous)
- Lou Gehrig's disease
- Macular degeneration
- Multiple sclerosis, lupus erythematosus
- Myocardial and coronary insufficiency, atherosclerotic plaque, arterial spasms
- Osteoarthritis
- Osteoporosis
- Parkinson's disease
- Peripheral vascular insufficiency, intermittent claudication, aortic calsinosis
- Rheumatoid arthritis
- Schizophrenia
- Scleroderma, Raynaud's disease
- Senile dementia
- Varicose veins

So Versatile - How Come No Misconceptions?

We assume that our readers are not only erudite but also inquisitive and eager to be given facts. In the context of Indian cultural attitudes and beliefs—from the medical standpoint—a majority of us seem to suffer from and also have a firm conviction about this peculiar "the doctor is always right" syndrome! You couldn't be more *wrong*! In the developed countries (Europe, USA, Canada, Japan and Australia etc.) today's patients adopt a more aggressive stance in dealing with their physicians. They are no longer reluctant to challenge their doctors and much prefer being dealt with on an equal footing. People today are generally more sophisticated medically, and growing numbers now approach the medical marketplace as more wary consumers. When a practitioner now recommends a treatment, the patient not only wants definite clarifications on what he's buying, but is also more apt to double and triple check as well! Unfortunately, even after 50 years of independence, the majority of us living in India still remain illiterate, uninformed, or just plain lazy and are prone to becoming the prime targets of exploitation—***especially in the medical arena.*** These pages are intended as a pathway to righting such wrongs. We want you **NOT** to visualize your doctor as some kind of god and insist that you exercise an active voice in matters concerning your personal well-being, care and cure. The only way to make this possible is to *learn.* No matter how much knowledge you have about cardiovascular disease, or even an above average comprehension of the circulatory system, be assured that you will still fall woefully short in making an informed treatment decision. Since chelation specifically targets heart ailments and age-related diseases, we're going to mete out information accordingly. Remember, atherosclerosis is *not* that simple a subject.

It would be best to start off by clearing up some of the more common, though readily accepted, misconceptions about chelation therapy. The reason why these repeatedly crop up, especially where the average Indian is concerned, is that most medical specialists are wont to display a snobbish attitude towards their patients and generally treat them like nincompoops. Physicians *do not* take the time to properly and fully explain a particular heart ailment; ***nor*** do they exercise the required patience when dealing with ordinary folk. We hope that the following clarifications will bring an awareness of the ground realities and prevent you from taking everything for granted.

Breathing difficulties, chest pains or leg cramps are invariably associated with atherosclerosis. Unadulterated bunkum! On the one hand you may experience chest pains and leg cramps and ***not have*** even an iota of heart disease. On the other hand you may already be a victim of advanced atherosclerosis ***without any*** of these expected symptoms! Though the most general indications of atherosclerosis and coronary artery disease are exertion-related chest pains, leg cramps or shortness of breath, there are many other non-cardiac conditions that can also cause chest pains. Three ailments where the pain caused is remarkably similar to that of heart disease are **gallbladder disease**, **hiatal hernia** and **shingles.** A **cardiospasm,** which results from an irritation to part of the **esophagus**, is sometimes not too different from true angina. Mild rib inflammation or arthritis in the joints between the ribs and spine or the joint between the ribs and breastbone can bring about chest pains. When you suffer a neck strain or whiplash injury, you contract a condition called "nerve root syndrome" which causes chest pain as well as an angina-like pain in the arms. Chest pains can also result from food poisoning or sleeping in an unnatural position or excessive air swallowing. **Aerophagics** (swallowers of air), much like people who hyperventilate or have over-breathing habits, are often the victims of anxiety and emotional tension. **Aerophagia** and **hyperventilation** may produce symptoms similar to those of serious circulatory ailments. Depression, a nervous upset or other psychological traumas can intensify symptoms that tend to imitate heart disease.

Many individuals experience a ***silent*** heart attack which may often be mistaken for just a bad case of indigestion or maybe an arm/neck muscle cramp that has inadvertently developed, until the next routine physical exam uncovers a completely different and far more deadly ball game! At times these folks do not recollect having had any symptoms at all. Typical indications of circulatory problems that people frequently do ***not*** associate with atherosclerosis are coldness of fingers and toes, or the gradual or transient loss of memory or impotence. **NEVER** come to your ***own*** conclusions that you, or some other family member, have actually developed atherosclerosis or any other heart problem. If you ever get edgy or harbor any doubts about your cardiac functioning, **DO NOT** assume that it is a heart disease or that it is not. Special tests can positively determine the presence, as well as the possible causes, of coronary artery disease. ***Get these done as soon as possible.***

The primary cause of most heart attacks is blocked circulation due to atherosclerotic plaque. Another load of tripe! Although mechanical blockage is one important cause, the degrees of arterial blockage have never been completely correlated with either heart attacks or the symptoms of coronary artery disease. Heart attacks, or **myocardial infarctions,** may be set off by either a mechanical blockage, that causes a complete cut off of the heart muscle's blood supply, or by a coronary artery spasm (which also causes sudden blood stoppage), or by both events occurring simultaneously. In both cases, malfunction results from disrupted blood flow and compromised oxygen supply. The muscle cells get damaged and an irreversible **infarct** forms resulting in a scarred and non-functioning heart muscle. Although angina pains, sometimes leading to a myocardial infarction, are now considered to often be the result of a spasm superimposed on pre-existing partial plaque blockage, heart attacks have been documented to occur in the ***complete absence of plaque and with perfectly normal arteries!***

It is an established medical fact that, *even in the complete absence of atherosclerosis*, coronary artery spasm or primary metabolic failure of the **myocardium** (heart muscle) can trigger a heart attack. There are numerous medical reports, and a surprisingly significant number of documented cases, to support the conclusion that ***spasm without atherosclerosis is not at all unusual***! In a complete turn about, there are also numerous medical reports of certain 'champions' with irrevocable evidence of complete blockage of ***all three main coronary arteries*** but who displayed ***no abnormal heart symptoms at all!*** They had experienced absolutely ***no*** angina and also showed excellent tolerance to strenuous physical exercise. The best illustrations of this fact are the autopsies carried out on the three astronauts who were killed in the tragic Project Apollo launch pad fire on January 27, 1967. All three of them, men in supposedly superb health, showed evidence of atherosclerotic disease – *one being such an advanced case that he would have been a prime candidate for bypass surgery had he lived and ever experienced the symptoms!*

When **prostacyclin** production is shut down by free radical activity, an unopposed spasm of blood vessels and an increase in platelet accumulation is caused resulting in clotting. This explains why angina or myocardial infarction frequently occurs after a meal that is full of peroxidised fatty acids that cause an explosion and flurry of free radical activity. It has become increasingly evident that the missing link is free radical pathology. While many will be familiar with the dangers posed to the heart by plaque accumulations inside the arterial walls, very few people actually know that the *equal or greater danger from arterial spasm is an irrefutable **fact.*** More alarmingly, not only is there a general lack of public awareness as to what causes arterial spasm but any knowledge of the preventive measures that might ward it off is also conspicuous by its absence! An improper calcium/magnesium ratio, within the arterial muscle, is an important factor in the cause of arterial spasm. If inter-cellular magnesium is low and calcium high, the heart muscle is unable to pump either rhythmically or adequately. The proper ratio of calcium ***outside*** cells to magnesium ***inside*** cells ***must*** exist if proper contraction and relaxation of muscle cells is to occur. This fact holds well for both the heart pumping muscles and for the muscle fibres that encircle the arteries and control blood flow.

Hardening of arteries is a natural physiological consequence of ageing. Stuff and nonsense! There is this mistaken conception that one's arteries are fated to harden, become rigid, finally get blocked and become inflexible as one ages. While it may be more prevalent in the elderly, hardening of arteries results from biochemical changes that may ***actually have started early in life.*** Free radical damage, which causes the formation of benign arterial tumors, or **atheromas**, and leads to hardening through cross-linkage and deposition of calcium, is directly linked with, and can be traced to, the high fat and highly refined diets of the industrialized world.

Low-cholesterol diets reduce the risk of atherosclerosis. In essence, not completely true. The statement would hold good, however, if the ***total*** amount of fats consumed is reduced, ***not just the cholesterol content.*** A major percentage of cholesterol in the blood is ***not*** a result of dietary habits but is **that which is produced by the body itself**. Actually, you will gain nothing by cutting out the oral intake of cholesterol because the less you consume *that much more is produced by the liver!* The main benefit of maintaining a low-cholesterol diet is the fact that the natural outcome is less consumption of total fat. Those unfortunates who suffer from high serum cholesterol, prime targets of atherosclerosis, endanger themselves *by eating too much fat* and ***not***, as is widely believed, from consuming high-cholesterol foods. So it is not cholesterol per se that is the culprit. The now popular dietary recommendation to consume less cholesterol may be good advice, but for all the wrong reasons! This is especially applicable to a country like

India where the majority of the populace have absolutely no ideas regarding dietary controls and ***definitely*** **NO** ideas about what's to be consumed as "**good for**" and what's to be avoided as "**bad for**".

Some physicians advocate that substituting margarine for butter and giving up eggs are effective ways of reducing cholesterol levels. It is claptrap to assume that a low-cholesterol diet lowers blood cholesterol when, as the foregoing explains, it is actually the reduction in harmful fats that turns the trick. Cholesterol is essential for many vital functions such as the emulsification, digestion and absorption of fats; the synthesis of sex hormones; the production of vitamin D via an interaction with sunlight; and the protection of part of the lipid membrane surrounding every cell. Food is not the only media for cholesterol to enter the body as it is also manufactured by the liver and other organs. Two-thirds of all the body's cholesterol is normally produced by its own cells, with the remainder resulting from dietary intake and other sources. Recent research has discovered that much of the cholesterol in deposits of atherosclerotic plaque is produced within the body and not directly derived from food. Some is even produced within the plaque itself! The fact of the matter is, that a majority of people can safely ***add*** one egg a day to their diet with no increase whatsoever in blood cholesterol—provided, of course, that those eggs are not fried in fat or scrambled in butter! It seems clear from experimental evidence that diets which contain less fat, as well as less cholesterol, can be of benefit and ***do*** result in lower blood cholesterol—thereby lessening the risk of contracting atherosclerosis.

Participation in vigorous sports and strenuous physical activity provides protection against the development of atherosclerosis. Again untrue. You'd probably collapse in shock to learn that autopsies carried out on the bodies of some highly active people, who had died of causes ***not*** related to any heart ailments, evinced atherosclerosis in highly advanced stages along with extensive arterial plaque buildup. And these people had never ever exhibited any symptoms of disease! Some of them were regular marathon runners in the prime of their lives! Similar research carried out on the nomadic Masai tribes in Africa, who routinely walk 20 to 30 miles a day, has also uncovered a surprisingly high incidence of atherosclerosis in a section of people long thought to be totally free of this disease.

But hold your horses! We are **NOT** suggesting that regular exercise is without any benefits or that devotion to an exercise regime is a waste of time. The physically active lifestyle of the Masai is a good example of the protective effects of exercise. Like other hyper-active people, while not immune to plaque buildup, the Masai rarely suffer symptoms of arterial blockage from this condition. Physical exercise does have a protective effect, preventing arterial occlusion by promoting collateral circulation around the blocked arteries or by causing compensatory expansion of the plaque-clogged arteries. While there is no proof that regular exercising will be instrumental in stopping plaque buildup, it most certainly *does* improve the ratio of HDL (**high density lipoproteins**) cholesterol to total cholesterol. In other words, the HDL percentage (the 'good guys' portion) increases in relation to total cholesterol, which ***will*** have a retardation effect on disease development. Individuals who maintain high levels of physical activity have many fewer symptoms of atherosclerosis and fewer atherosclerotic-related deaths, even though many of them may in fact have extensive arterial plaque. New research also suggests that vigorous exercise reduces the risk of 'sudden' death from coronary heart disease. A University of Washington research team had evaluated the activity levels of 1250 persons who had died suddenly and unexpectedly of heart disease in 1980. It was discovered that these sudden-death victims had led sedentary lifestyles, almost totally devoid of any high-intensity exercise. In comparison, a matched group that was far more active showed a much lower percentage rate of sudden-death occurrences. Another study revealed that bus drivers, who did all their work sitting down, had a much higher heart attack rate than did bus conductors whose work involved much more exercise.

Though exercising will certainly help protect one from the harmful effects of arterial disease, however, it *does not* ***prevent*** the disease from developing—which initiates the groundwork for the harm it will eventually cause. The only possible conjecture that we can arrive at is, perhaps, that in the case of physically active people this eventuality will occur in old age rather than middle age! Unlike the West and some other countries, the benefits of exercise and the ultimate goals are privy to only a fortunate few over here. In our various towns and cities, so-called health parlours have mushroomed as a result of television advertising hype and are more a money-making racket rather than a health asset. In some cases, the practices employed are dangerous enough to cripple a person for life. Even so, meaningful legislation governing the proliferation of these and their practices is yet to be enacted. You are advised to exercise extreme caution when approaching one of these for the touted "health benefits". What needs to be remembered is that exercising *itself* is ***not*** the be all and end all therapy.

Hardening of the arteries begins with the buildup of calcified fatty deposits, called plaque, along the inside wall of coronary arteries. Definitely not. The onset of arterial disease begins with the damage to arterial lining resulting from blood flow stress, routine wear and tear or from free radical damage. If you are to understand the basic concepts underlying the development of atherosclerosis you will have to use your imagination realistically. Close your eyes and try to visualize a cell in the arterial wall that is in the process of multiplying abnormally. The growth gets bigger and bigger until a tumor forms. Though this tumor is non-malignant and does not spread to other parts of the body, it ***has***, nonetheless, formed an unwelcome, space-occupying mass on the inside of the artery. Other connective tissue constituents (collagen, elastin, etc.) start accumulating on it making it bigger still. When this growth exceeds its blood supply of oxygen and nutrients, it starts to break down at its centre, commences to decay and gradually starts gathering deposits of calcium and cholesterol. As it grows into what we call **plaque**, it progressively gets firmer and more rigid. This should make you realize *that calcification is actually a late occurrence in plaque formation*, not the initial event. These growths are akin to the cellular proliferation seen in cancer. The arterial wall cells mutate in response to free radical damage to the genes in the nucleus, exactly in the same manner that atomic radiation causes mutations. An occasional cell loses its ability to control cell division and multiplication, resulting in the tumor-like growth also called an **atheroma**.

Chelation removes abnormal deposits of calcium from arterial plaque. Totally untrue. Before the free radical causes of degenerative diseases were known, it was generally assumed that EDTA chelation had its major effect on calcium metabolism. We now know that calcium is just another link in the chain of cause and effect created by free radical damage. Although EDTA is certainly a calcium chelator, it has far more affinity for other metals (iron, copper, lead, mercury, etc.) and will quickly drop calcium to pick up one of these.

Low Affinity for Calcium? That's of No Use, Then!

This major misconception about chelation therapy needs clearing up once and for all. Chelation does ***not*** *strip* calcium out of plaque or 'hardened' arteries. In the earliest days of the therapy, in their eagerness to offer simple and seemingly understandable explanations to their patients, some of the pioneers described chelation as an action akin to "removing the cement in the blood vessel linings". Though crude, this was a vivid way of describing what *they* thought was happening—that calcium was being dislodged and drained from the plaque-clogged and hardened arteries. EDTA's proven calcium-lowering ability coupled with the undeniable evidence of overall improvements in the health of patients, only strengthened the belief that led to the seemingly reasonable conclusion: the main job of EDTA was to 'zap' calcium out of the system! In its early days EDTA was generally referred to, whether in medical or industrial applications, for the ultimate 'removal' effects that it had on calcium. For the everyday consumer, its ability to remove calcium meant cleaner bathtubs and shirt collars! Since degenerative diseases have long been linked with calcium-related excesses,–even cardiologists and internists hold EDTA in high regard for this value knowing that it will lower calcium levels in patients having life-threatening high blood calcium. It is an established fact that, as people age, their inside mechanisms tend to deposit and accumulate calcium in the most unwanted places, especially in the arteries and already formed arterial plaque.

As stated previously, EDTA's affinity for calcium is far less than its attraction towards other metals and it will discard calcium to pick up these instead. On the flip side, it is a scientifically supported fact that there truly *is* a calcium-chelation connection, but this is secondary to EDTA's impact on *free radicals*. It is actually the free radical-chelation connection that is most responsible for the positive improvements in individuals with disorderly calcium metabolism, atherosclerosis and all degenerative diseases. Therefore, the next question to logically ask is, "If EDTA does not simply 'pull out rivets' or 'decalcify plaque', than how on earth does it reverse the symptoms of arterial disease?" We have already learned how arteries start to harden, how 'injuries' to the walls occur leading to scarring and depositions and how the growth or atheroma starts. We have also seen how the central core of the plaque degenerates and starts collecting calcium ***and*** cholesterol deposits and increases in size till it effectively blocks blood flow in the vessel.

Calcification is a relatively late event in the development of both **arteriosclerosis** (the hardening of arteries) and **atherosclerosis** (the hardening of the growth or plaque). While there are numerous nutritional and lifestyle aspects that contribute directly to the formation of atherosclerotic plaque and its calcification, we are going to specifically deal here with that form of disordered calcium metabolism that is most directly corrected by EDTA. Long before abnormal calcium deposits become solidified and actually visible, a progressive fine mist of calcium is building up within cells and connective tissues. Abnormalities first occur in relationships between calcium concentrations such as *bone versus soft tissue* and in the *ratio of calcium to magnesium*. So, the absolute amount of calcium in the body is NOT the most important factor. An optimal calcium-to-magnesium ratio results in a much higher concentration of calcium ***outside the cells*** than within! Healthy cells, with active energetic metabolism, are able to maintain an efficient pumping system within their lipid membrane walls that keeps calcium out and magnesium in. With ageing, this concentration gradient decreases, with calcium levels increasing within the cells as the magnesium levels decrease. This is partly because of the gradual breakdown of the cell's pumping mechanism, which can be damaged by many things, ***including free radicals***. Free radical damage to the cell wall can create a leak that permits calcium to seep in and allows magnesium to seep out. The more calcium that leaks into the cell, the more poisoned its metabolism becomes, with eventual crystalline deposits forming inside the cell and causing it to die. The more cells that die, that many fewer are left to continue with the life process.

That Still Means a Calcium – vs – EDTA War, Doesn't It?

No, not really. Excess intracellular calcium and abnormally high calcium-magnesium ratios speed up cell death, especially when blood flow and oxygen delivery become compromised by atherosclerosis. Arterial spasms become much more intense and calcium leakage into cells proceeds even more rapidly once the cellular calcium-magnesium ratio is disrupted. A vicious cycle sets in as diminished blood flow and reduced oxygen supply speed calcium leakage into the cells. Intracellular calcium causes the muscle cells surrounding arteries to go into more intense spasm causing further reduction in blood flow and oxygenation, which speeds up calcium influx – and so on and so forth! In addition, a faulty calcium-magnesium pump allows more ionized calcium to enter the cell activating an enzyme that leads to the production of prostaglandin-related **leukotrienes** (a chemical reaction that releases free radicals) which are potent inflammatory substances that attract white blood cells to the area. When excessively stimulated by leukotrienes, WBCs tend to run amok and initiate excess free radical production which, in turn, causes increasing inflammatory damage to healthy tissues. Small blood vessels dilate causing leakage of red blood cells and platelets through the blood vessel walls, causing microscopic clots to form. Some RBCs then hemolyze, releasing copper and iron which in turn catalyze an increase of free radical destruction to lipid membranes in the vicinity, triggering another vicious circle. Once free radical production has exceeded the body's capacity to protect itself, metabolic breakdown, cellular damage and tissue calcification all take place more rapidly. Calcium deposition is just another link in the chain of 'cause and effect' created by free radical damage.

EDTA *does* influence and speed up the removal of calcium deposits in many ways as part of the healing process, but not as directly as it was originally imagined. However, it ***does reduce*** blood calcium levels during infusion and hastens excretion. However, if one were to consider that a total of just 1500 grams of calcium exists in the body, then the one-third of a gram excreted is so relatively small that it couldn't possibly be the main benefit of the therapy. During the short time that EDTA is circulating in the body (it has a half-life of only one hour), it temporarily lowers blood calcium. The resulting drop in serum calcium provides a stimulus to the **parathyroid gland** to step up production of **parathormone**. This hormone, in turn, signals **osteoblasts** in the bone to increase their normal bone calcification production. A three-month waiting period, after the completion of a course of EDTA infusions, is normal before the benefits become obvious. This effect of EDTA is actually its ability to reduce the rate of pathological free radical chemical reactions by removing toxic heavy metals and restoring metabolic health to a level from which the body's own defences can now take over, thereby providing time for free radical damage to be repaired by natural healing.

Degenerative? Look Out! Look Out!

There are some diseases which could result because of the deterioration and loss of specialized functions of the cells of a tissue or organ which drastically alters the character of the tissue or organ. The changes may be due to a defective blood supply or involve the deposition of calcium salts, or fat, or fibrous material in the affected organ or tissue. Such changes in character and function are considered a degenerative disease. We will now explain some of the degenerative diseases, caused by lack of free blood flow through the vessels, that respond to chelation therapy.

Diabetes. This is one of the disease reactions to poor circulation. It is quite rampant in India and, next to heart disease, is probably the second most common cause of debilitation and death in that country. Diabetics are prone to having their blood vessels clog with fat and 98 percent of the victims also suffer with diabetes-influenced complications such as cataracts, heart disease, kidney damage, blindness, blood vessel damage, nerve disorders and obesity. Diabetic patients are treated with a course of chelation therapy in such a manner that they are brought back to eating more unprocessed foods. Certain foods are natural chelators. These are bran, whole-grain, wheat, oats, corn, cereals, lentils, beans, peas, peanuts, figs, dates, apples and other rich sources of fibre and complex carbohydrates. The mineral chromium, in the form of a glucose tolerance factor, is another natural chelator for diabetics. Diabetics can also have their insulin requirements reduced by EDTA chelation.

Gallstones. In this case, a normally soluble component of bile separates and grows in the gallbladder producing a predisposition to disease. Eighty five percent of all gallstones are composed of cholesterol—whose concentration in the blood is reduced by chelation therapy. Besides lowering serum cholesterol, therefore, EDTA also dissolves gallstones.

Kidney Diseases. Those that are related to circulation difficulties usually arise from the narrowing of the renal artery, which is divided into smaller branches flowing through a kidney and its filtering units. Kidney stones get dissolved by chelation and the therapy, far from hurting the kidneys—even already impaired ones—is known to improve kidney function. However, since all heavy metals themselves are a source of kidney poisoning, particularly when the heavy metal is being eliminated through the kidney, some precautions have necessarily to be taken for kidney patients. This aspect will be discussed separately.

Clots in the Blood Vessels. These can bring about strokes, phlebitis, pulmonary embolism or varicose veins. Certain antioxidant nutrients, such as vitamin C, vitamin E and selenium, also happen to be chelating agents and are often recommended by chelating doctors as oral chelators for patients with blood clots. Those who suffer from a blood clot problem called **thrombophlebitis**, will have **heparin** added to the EDTA solution during intravenous infusions.

Emphysema. The patient has microscopic air sacs of the lungs that have lost their elasticity. Air flows into them, but cannot flow out easily because of the narrowed sac passageway. As this trapped air cannot be expelled, the patient retains too much stale air in the lungs. Like tiny balloons, air bubbles take up space in the air sacs stretching the thin walls to the rupturing point. Gas exchange is reduced. The situation leads to shortness of breath, over-working the heart and possible death. Chelation helps restore blood transport and good nutrition, which is mandatory for this lung condition.

Cataracts. These are solidified forms of a person's normally transparent lenses, producing an opacity whose curtain-like action prevents light rays from reaching the retina, the eye nerve and the brain. Hardening of the arteries causes a partial blockage of the blood supply going to the eyes and is frequently responsible for the cataracts. Cholesterol deposits are sometimes another cause. Chelation therapy is a specific adjunctive way to bring about a biochemical change for improvement of the eyes without resorting to cataract surgery.

Senility and Alzheimer's Disease. This could involve atrophy of brain cells from cholesterol-filled plaques which will, naturally, narrow blood vessels in the brain. Another manifestation of senility is Alzheimer's disease, a relentlessly progressive form of senile dementia, which hits about 5 to 10 percent of the world's 'over sixty-five' population. With Alzheimer's, the brain's aluminum concentrations are four times higher than normal, causing structural damage of nerve **plaques** and **tangles**. Plaques are hard knots of debris marking the site of a burned-out nerve cell ending. Tangles are clumps of filaments that form within the nerve cell body. Chelation therapy is the ***only*** known treatment available that successfully combats Alzheimer's disease.

Macular Degeneration. In this arteriosclerotic disease, blood is prevented from reaching the central part of the retina through the **choroid** capillaries. A blind spot develops in that part of the vision that is used for focusing on an object. The patient will not be able to read, will suffer from colour blindness and not be able to recognise others. Chelation therapy has proved its effectiveness in restoring circulation and improving vision in these cases. It is better than any other therapy, even eye surgery, currently available to tackle this dysfunction.

Varicose Veins. These veins in the leg possess damaged valves that are supposed to lift the blood against gravity, prevent backflow and then return the blood to the heart. The causes of damage could be bacterial infection, obesity, injuries that result in blood flow obstruction, poorly oxygenated venous blood or metabolic waste products in the bloodstream. The problems can be overcome by enhancing vein circulation through increasing the blood nutrition to the veins themselves. This is accomplished with either oral or intravenous chelation treatment.

Parkinson's Disease. This is a slow, progressive condition of the **basal ganglia** located in the small centre at the base of the brain. It does not affect the major portion of the brain, allowing the patient to retain intelligence, not suffer any paralysis and also retain the ability to speak. The usual indications are tremors, stiffness, shuffling gait and slowness of movement. The response to treatment by chelation therapy is positive and good, especially in the beginning stages of the disease.

Osteoporosis. This is the pronounced porosity of bones that frequently occurs in menopausal women. It results in sudden bone fractures for no apparent reason. Bone re-calcification can be stimulated by continued **osteoblastic** activity (osteoblasts are the body's bone-forming cells) which uptakes calcium from blood serum as a result of EDTA intravenous administration. With osteoblastic activity, the soft tissue pathologic calcium in plaques continues to diminish in order to meet the need caused by this increased bone uptake of calcium. Serum calcium is continuously leeched back into bones by activated osteoblasts, resulting in serum calcium being necessarily replenished from other body stores. This restoration is accomplished by removing some of the calcium from soft tissues and pathological stores, such as from arteries and around joints. Hardened arteries soften and widen while arthritic joints get more mobile and the pain stops. Moreover, the bones and teeth get stronger because ionic and metastatic calcium, which may have eluded the grasp of the chelation 'claw', go into reinforcing the existing bone.

Paget's Disease. This is a chronic process of bone destruction, overgrowth and new bone formation which occurs in the elderly and most frequently affects the skull, backbone, pelvis and long bones. Also frequently, hardening of the arteries and impaired blood supply to the bones are clearly evident. Chelation therapy works to relieve the pain and deformity associated with this disease.

Hypertension. This is more of a symptom than a disease. It is actually an elevation of blood pressure far exceeding the normal body limits. Usually, hypertension is present when the systolic pressure is above 140 and the diastolic exceeds 90. One of the fastest benefits seen from chelation therapy is lowering of high blood pressure to normal.

Give As Good As You Get!

It is time to take stock and summarize the 'goodness' of chelation therapy by considering some of the physiologic effects of EDTA in relation to free radical pathology. By now you should be more than just familiar with all the relevant aspects of disease and treatment. We're just helping you to move forward by putting things in a nutshell (some nuts!). So, getting to specifics, we 'aver' with total conviction that:

- **EDTA** controls free radical damage. Free radicals cause cell destruction by initiating chain reactions of lipid peroxidation which disrupt fatty membranes within and surrounding cells. Lipid peroxidation requires the presence of abnormally located metal ions such as excess iron. EDTA removes these metal ions—which accumulate with age—and permits normal healing of peroxidative damage by stopping further peroxidation.

- **EDTA** removes lead and other poisonous metals restoring normal enzyme functioning. Excessive concentrations of heavy metals cause enzymatic dysfunctions independent of their action as free radical catalysts, with a concomitant loss of cellular homeostasis and failure of vital cell functioning.

- **EDTA** enhances the integrity of cell membranes. The ability of cells to produce and store energy depends on healthy membranes surrounding each cellular compartment, including each little **mitochondrion,** the tiny energy factory in the interior. It also stabilizes mitochondrion membranes and enhances the efficiency of energy metabolism independent of any effects on blood flow or oxygen.

- **EDTA** helps reestablish **prostaglandin** hormone balance which in turn prevents arterial spasm, blood clots, plaque formation and arthritis. Prostaglandins are extremely potent hormones, having a half-life measured in mere seconds, and must be constantly synthesized. The two most important prostaglandins, in relation to blood vessels and atherosclerosis, are **prostacyclin** and **thromboxane**. The former reduces the tendency of platelets to stick together, thereby causing blood clots, and also reverses blood vessel spasms. The latter does exactly the opposite. It causes intense spasm and stimulates platelets to become very sticky, making the blood vessel walls resemble cellophane tape. A proper balance between the two must be maintained to protect against injury and hemorrhage on the one hand, and to maintain normal circulation on the other. Prostaglandins are produced from fatty acids and production is put into imbalance by destructive lipid peroxidation. EDTA inhibits such lipid peroxidation by chelating out the catalyzing metallic ions.

- **EDTA** protects the integrity of blood platelets. Platelets are very small, semicircular corpuscles in the blood stream that quickly adhere to areas of injury and release substances that diminish blood loss by forming clots. In the process of clotting they change shape, tend to become too sticky and attach themselves too readily to the walls of diseased coronary, cerebral or other arteries. Platelets also release spasm-inducing **thromboxane**. When platelets come into contact with EDTA, their tendency towards over-coagulation is drastically reduced.

- **EDTA** sweeps up the minute particles (molecules) of lead, cadmium, mercury, iron, aluminum, and other metals deposited abnormally through the body. Concentrations of metallic ions that have the ability to catalyze free radical chain reactions are so tiny that even the immeasurable residue in distilled water may initiate such reactions. The addition of EDTA completely blocks these free radical chain reactions that would otherwise occur by binding metallic ionic catalysts, making them inert and removing them from the body.

- **EDTA** normalizes calcium metabolism by reactivation of the enzymes poisoned by lead and other heavy metals. Most importantly, it neutralizes free radical activity which converts cholesterol into substances with vitamin D activity, activity which causes plaques to accumulate calcium.

- **EDTA** intermittently lowers serum calcium, allowing gradual reversal of abnormal calcification by stimulating the uptake of calcium in the bones.

- **EDTA** increases tissue flexibility by uncoupling age-related cross-linkages. Cross connections between large protein and other connective tissues' molecules cause the increasing rigidity and loss of flexibility so common with ageing. Cross-linkages to enzyme proteins can interfere with normal biochemical activity.

- **EDTA** encourages overall metabolic efficiency due to the whole range of all these factors.

It's Really Got What It Takes!

The development of atherosclerosis, or any other degenerative disease, takes decades and is almost always the result of not one, but a combination of health-destroying factors. Similarly, reversing the process also ***cannot*** rely on just one treatment session because impaired health can only be restored by therapies that counteract the entire spectrum of disease-causing factors. Underlying every known risk factor contributing to degenerative diseases, from atherosclerosis to cancer, ***is the excess free radical activity with resulting free radical pathology***. Improper diet causes free radical damage as do smoking, alcohol and drug abuse, insufficient exercise and excessive stress. Add to these the lack of adequate vitamins, minerals and trace elements and you have a potboiler that defies definition—let alone reasonable diagnosis! The key to restoring health is to reverse the free radical pathology gradually by adopting health-promoting lifestyle strategies and, if the need so arises, **MORE DIRECTLY AND QUICKLY BY A COURSE OF CHELATION TREATMENTS.** With your newly acquired understanding of free radical chemical reactions, we are certain that you can appreciate the efficacy and long-term benefits of this therapy.

Absolutely Marvelous! But – No Risks?

Before any person embarks on the chelation journey, it is essential that a detailed health history be compiled ***and*** a complete physical examination be performed. This would include all the specialized investigative tests if a patient is already in the grip of some disease. On completion of all clinical tests, an analysis will be made and a viable treatment plan drawn up. EDTA infusions, when compared with other types of treatment, are non-toxic and practically free of dangerous side effects. Presently, the risks of having some deadly reaction to the treatment are about 1:1,000,000 – and 'deadly reaction' is not to be misconstrued as 'death'. Absolutely ***nobody*** dies from chelation therapy! In complete contrast, facts reveal that three out of every 100 patients die when undergoing major vascular surgery. Unfortunately, this percentage is the result of occupational hazards and remains static (may even increase) despite the efficiency of the surgical team even being 500 percent! Naturally, the complications arising from any type of surgery—heart failure, strokes, infarcts, bacterial infections—are certainly much higher than the 'no risks' of chelation therapy. We're not even going to mention post-surgical pain. Rest assured that chelation therapy is not only a million times safer than any surgical procedure *but is also completely pain free!*

Of course, a treatment that is as effective as chelation therapy will also have some side effects. These are easy to detect and equally easy to control. Though these effects are extremely minimal and occur usually during the initial phase of treatment, we mention them because of ***your need to know***. The following is a list of the possible side effects of **EDTA** infusion:

- A headache or slight shivering or dizziness after the infusion.
- Bloating.
- Acid indigestion or even an upset stomach.
- Burning sensation or black and blue mark at the infusion site.
- Back pain.
- Skin rash.
- Tiredness or a feeling of weakness.
- Adverse kidney reaction (will be discussed in detail).

All the above side effects slowly disappear during the course of the therapy. Seldom is it necessary to stop the treatment. The regimen is almost as simple and safe as taking an aspirin tablet—assuming, of course, that you're not allergic to aspirin! Patients have no perambulatory problems and no weakness or loss of any senses after treatment and are able to move 'on their own steam' without difficulty. Thus the therapy is strictly an ambulant treatment that can be safely performed almost anywhere ***with no hospitalization or physical incapacitation involved!*** However, just as there would be side effects of medicinal over-dosing, if EDTA is infused too rapidly or in too high a dose, there could be some serious side effects. Too rapid a chelation infusion will cause a marked drop in the blood's content of calcium ions. Known as **hypocalcemia**, it is the condition of insufficient calcium ions in the blood. While mild hypocalcemia has no symptoms, severe hypocalcemia can bring about irregular heartbeat, muscle spasms, and a burning or prickling feeling of the hands, feet, lips and tongue. These side effects are easily corrected by either stopping the IV drip or decreasing the rate of infusion.

But What About the Kidneys?

PLEASE DEVOTE A LITTLE EXTRA TIME FOR THE PERUSAL OF THESE IMPORTANT OBSERVATIONS AND FINDINGS. In the treatment of lead poisoning with EDTA, it is possible to experience kidney failure. This may happen not because of any toxicity of EDTA, but because of its powerful chelative properties and the fact that the 'bound up' lead is eventually excreted through the kidneys, which are prone to being poisoned during its passage. An excessive body burden of lead already present in a patient, without having been detected prior to the start of chelation, could cause this lead nephrotoxicity. The lead is apt to poison those tiny filtering units called nephrons. These resemble tiny funnels with long stems, each having two twisted tubes. Urine is formed in the renal corpuscles, and in the tubes, by the process of filtration, re-absorption and release. The collecting tubes carry the urine to the kidney, pelvic area and ureters. The excess amount of lead, floating in the blood and eventually 'captured' by EDTA, may result in the poisoning of the kidneys at the moment of excretion, ***especially if the kidneys have suffered any sort of impairment in the first place.***

Contrary to the above, EDTA ***does not precipitate any damage*** to kidneys that are functioning normally. In fact, the therapy enhances kidney performance. There is enough clinical evidence to support this and there is no cause for undue alarm in this regard. What should be ensured is that all the pertinent tests for kidney function are thoroughly done immediately preceding the therapy. This will also help in determining the exact protocol for each individual. The same tests must be repeated immediately after every third chelation session. This will present a clear indication of EDTA's ultimate effects and also, if needed, dictate any requirements for re-tailoring. As the therapy sessions progress, you will find that kidney functions improve in leaps and bounds, thereby permitting the infusion of larger doses without any debilitating effects.

Whew! The Contra-Indications Continue!

When a vein gets irritated due to the repeated insertions of a needle, or for any other reason, there is a possibility of an inflammation, with clot formation, being induced at the site. This phenomenon is known as **thrombophlebitis**. Generally, this is ***not*** the kind of clot which is likely to break away and travel through the circulatory system causing blockages and other problems. It is more of a local inflammatory response to irritation which the body will take care of by itself over a short period. Thrombophlebitis may be annoying, but it poses no serious consequences for any patient and it can occur during any type of infusion—not specifically just because of chelation.

A drop in blood sugar, or **hypoglycemia,** is another side effect of EDTA infusion. The symptoms are headache, problems of vision, weakness, loss of muscle coordination and anxiety. The antidote is simple; swallowing a glass of fruit juice prevents or offsets the blood sugar reduction. Simply by eating fruits, sandwiches or any other food ***while undergoing chelation therapy***, patients do away with the entire problem of possibly becoming hypoglycemic.

Patients with a history of congestive heart failure, or those who suffer from high blood pressure, need special monitoring during chelation therapy. The sodium contents of the IV solution need to be reduced in such cases with the avoidance of saline, Ringer's lactate and other sodium-containing mediums. The fluid-electrolyte balance must be carefully controlled. Potassium and manganese depletion must be prevented. The infusion timing should be slowed considerably and the solution amount reduced almost by half.

At the same time that EDTA removes heavy metals and other toxic minerals, because of its efficiency as a chelator, it depletes some of the essential nutritional trace elements. The replacement of these depleted essential trace elements, is absolutely necessary. The amount of losses, or the percentages of successful replenishment, can be determined through urine measurement, hair mineral analysis and blood examination that should be carried out prior to and immediately following chelation sessions.

There has always been the fear that the treatment of arterial plaque with EDTA would weaken the walls of the arteries. This is a totally wrong assumption because it has been definitely proved that EDTA infusions restore the arteries to their full elasticity and strength. It is also a fact that EDTA does ***not*** cause decalcification of the bones, cartilage or teeth. In quite the opposite reaction, due to an increase in the secretions of the parathyroid gland, significant improvements in the metabolism of bone result.

An occasional patient may develop a skin rash, either as an allergic reaction or directly related to the body's loss of certain vitamins (such as B6) and minerals (such as zinc), even though these substances are supplemented during treatment. When found necessary, more amounts are added. Then again, the therapy may suddenly bring about an upset tummy or even induce some nausea. These effects are experienced by less than one percent of all patients and are readily treated by medication or increased vitamin B6. If you happen to get a headache, a simple pain-reliever does the trick. Another one percent of patients tend to develop diarrhea, or a loose stool, either immediately after the therapy is over or on the next day. This is a problem that is easily remedied by any anti-diarrhea medicine. More common is the need to empty the bladder more frequently. Some patients have reported getting up more than twice at night to urinate. Since EDTA is a mild diuretic, the therapy improves tissue integrity, decreasing soft tissue edema and returning the accumulated fluids to the circulatory system for final excretion. This is the reason why many patients reap the unexpected bonus of experiencing a beneficial weight loss as well.

Once in a while, in extremely rare cases, there may occur sudden feelings of faintness, weakness, light-headedness, dizziness or extreme fatigue on standing up suddenly, either during or after treatment. This could happen because the blood pressure does lower during the course of the therapy. It is an established fact that, after chelation therapy, the dosages of anti-hypertension medication have been drastically reduced in many cases of those suffering from high BP. Resting for an extra hour or so after treatment with feet elevated and the head lowered till the BP has normalized, can immediately rectify the problems of dizziness etc. Sometimes there is an individual case of unexplained fever within 24 hours of being chelated. This usually disappears spontaneously without any additional treatment but should be brought to the notice of the attending physician.

The foregoing information is given to enable a better understanding and awareness of some of the limitations of chelation therapy as well as providing an appreciation that the overall beneficial effects far outweigh all negative aspects. An important point to realize is that nowhere has anything been advised **AGAINST** chelation therapy. The applications of EDTA are almost limitless in their versatility and chelation has positively assisted even those with "one foot already in the grave"! EDTA chelation can be administered to anyone, regardless of age or the severity and extent of a resident disease. All that your chelation doctor has to do is to correctly calculate the compilation, dosage and rate of infusion of the IV solution for each particular individual, keeping in mind the type of dysfunction and extent of debility and whether prescribing an oral chelation would give similar results. If all the necessary precautions are recognised and carefully exercised, the resulting post-therapy exhilaration is all that you really have to look forward to!

If It's So Good, Why Is It So 'Unknown'?

There are a good many reasons for chelation therapy, and practically all the other new age therapies and medicines, not being as popular or as well known. However, since this aspect requires a bit of detail, we will discuss it in a dedicated chapter a little further on. For the time being, suffice it to know that, even though the 'good news' is slightly slow in percolating, at the present time chelation therapy has a fairly wide international following. Many orthodox practitioners are finally beginning to see the light in the understanding and appreciation of the value and long-term benefits of this treatment. As a result, hundreds of thousands of patients world-wide have received the therapy, with varying degrees of success, and the single uniting factor for this entire 'horde' is the fact that *all* of them have been highly pleased with the results, even in the many instances where the cures have only been partial! This certainly speaks volumes for its efficacy!

Switzerland, Germany, USA, Austria, France, Italy, Greece ***and*** India are some of the countries where it is now possible to obtain chelation therapy fairly easily and on a regular basis. This is because chelation solutions contain substances that have been approved by the health ministries of all these countries. Lack of general knowledge concerning the subject is one of the main reasons why we in India are lagging somewhat behind in what should otherwise be widespread local usage. It is hoped that this book will help you to develop a voracious appetite for knowledge about alternative therapies and medicines and turn things around for the better—sooner rather than later! It is quite understandable that the 'established' medical establishment must be presented with proof that chelation therapy is ***actually*** the viable alternative treatment for degenerative diseases including arteriosclerosis. Unfortunately, the massive amount of clinical evidence that has been collected over the years, proving beyond any doubt that it is a safe and effective treatment, has never appeared in any mainstream medical journals. Most journals had refused to publish any articles on chelation because of a threatened boycott by the pharmaceutical industry. That is one reason why many physicians are unable to get hold of these results. However, circumstances are definitely changing for the better these days as interested doctors can obtain whatever information they require directly from chelation clinics themselves. The other ray of sunshine is that loyal world-wide practitioners of chelation therapy are extremely hopeful because more and more regular general practitioners seem to be evincing a keen interest in the therapy. Though the orthodox medical profession has always been very conservative in and skeptical about accepting changes, several chelation doctors have reported that, of late, a fairly large number of patients have been referred to them by cardiologists!

The time is ripe for a fresh recognition and evaluation of chelation therapy in India. The fact of EDTA's successes in the treatment of cardiovascular disease, far from being just the icing on the cake, is actually the key that will open up the portals to a greater and better understanding of ageing and good health. It is certainly very foolish to dismiss chelation therapy as the viable alternative to invasive surgical procedures. Isn't it amazing that here we have, readily at hand, a marvelous therapy that ***cures*** over 90 percent of patients and, for purely conceited reasons, we refuse to exploit its entire potential. It is sad that chelation has no applications for gross stupidity! With time, however, the truth will definitely be emblazoned up front and any therapy or medicine that truly heals will, again with time, ultimately become sovereign.

Clinical benefits from chelation therapy vary with the total number of treatments received and the severity of the condition being treated. More than 75 percent of chelation patients have improved dramatically. More than 90 percent of patients who receive 35 or more treatments have benefited tremendously; even more so when and if they have also corrected dietary, exercise and smoking habits, which are known to aggravate occlusive arterial disease. Symptoms improve, blood flow to diseased organs increases, need for medication decreases and, most importantly, the quality of life becomes much more productive and enjoyable. When one first hears about or is seriously considering the option of trying out EDTA chelation therapy, there are normally a lot of questions that need answering and clarifications. Undoubtedly, you have some too!

Let's Start the Quiz Then!

Here are the most commonly asked questions with appropriate answers, explained as simply as possible in non-technical language. These are questions that you might ask out of curiosity or as a means of gaining some insight into what makes chelation tick.

What exactly is chelation as a medical therapy? Chelation is a treatment by which a small amino acid called EDTA is administered to a patient intravenously (IV), under the supervision of a licensed physician. The fluid containing EDTA is infused through a small needle placed in the vein. The infusion bonds with excess metals in the body and carries them away in the urine for final excretion. Abnormally situated nutritional metals, such as iron, along with toxic elements such as lead, mercury and aluminum are easily removed by this therapy. The percentages of those minerals and trace elements essential for health, which are normally present and more tightly bound within the body, can be maintained with a properly balanced nutritional supplement.

How many treatment sessions will be required? Chelation therapy is a course of treatments usually consisting of anywhere from 10 to 50 separate infusions, depending on each patient's individual health status and the severity of particular maladies. For patients with symptoms of arterial blockage, on an average about thirty sessions are required for the optimum benefits to accrue. Some patients with multiple diseases and complicated conditions may eventually receive even more than 100 treatments over several years. Each infusion lasts from between two and a half to four hours and could go some hours longer for the physically weak or those being treated for debilitating diseases that are present in an advanced stage. The frequency is normally one to three sittings each week. Over a period of time, EDTA will halt the progress of free radical activity and effect total cures. As you are already aware, free radicals underlie the development of atherosclerosis and many of the other degenerative diseases of ageing. Reducing the damage by free radicals allows diseased arteries to heal, restoring blood flow. With time, chelation therapy also brings about great improvements in many of the essential metabolic and physiologic functions within the body, besides restoring the regulation of calcium and cholesterol by normalizing the internal chemistry of cells. EDTA has beneficial effects on the entire flow of blood through every vessel in the body, from the largest arteries and veins to the tiniest capillaries and arterioles, most of which are far too tiny or too deeply located within the brain or elsewhere to even contemplate effective surgical treatment. In many patients, it is these smallest of blood vessels that are probably the most severely diseased. The healing benefits of chelation extend from the top of the head to the soles of the feet, covering the entire human physiological spectrum—not just in the short segments of the few large arteries which could, anyway, be bypassed through surgery!

Does it hurt? What does it feel like to be chelated? Being chelated is quite a different experience when compared with other medical treatments. There is no pain and, in most cases, very little discomfort. Patients are seated in reclining chairs and can read, nap, watch TV, do needlework, or chat with other patients while the EDTA solution is being infused into their bodies. If necessary, a person can even walk about, visit the restroom, eat and drink as desired, make telephone calls—all the while being very careful not to dislodge the needle that has been inserted into the vein. Some patients even run their businesses by telephone or computer while receiving the therapy!

What are the types of examinations and testing that must be done prior to beginning chelation therapy? Prior to commencing a course of EDTA therapy, a detailed medical history is essential. We are interested in knowing ***ALL*** about you; Not just your age, marital status, occupation and all previous medical problems, but also about your family and ***their*** health status. Your diet will be analyzed for nutritional adequacy and balance, and pertinent medical records and summaries of hospital admissions will be scrutinized. A thorough head-to-toe, hands-on physical examination will then be performed. The listing of current medications will include the frequencies of treatment and dosage strengths with special note being made of any allergies. Blood and urine specimens will be obtained for a battery of tests to ensure that there are no existing conditions which may be worsened or adversely affected by chelation. Kidney function will be carefully assessed. An electrocardiogram and chest x-ray may be ordered. Other non-invasive tests will be performed, as medically indicated, to determine the status of arterial blood flow prior to therapy. A consultation with other medical specialists may be requested.

Is it legal? Absolutely. There is no legal prohibition against a licensed medical doctor using chelation therapy for whatever conditions he or she deems to be correct. In India, the Health Ministry has neither disapproved nor rejected the use of EDTA. What a patient would need to establish is whether the physician carrying out the treatment has undergone the requisite and bonafide training in respect of chelation therapy.

What proof do you have that it works? Chelation physicians with extensive experience and practice have observed dramatic improvements in the vast majority of their patients. Angina has been routinely relieved; patients suffering from searing chest and leg pains and capable of only limited movement are frequently able to return to normal, productive and healthy living after having undergone the therapy. Far more dramatic, but an equally common occurrence, is seeing diabetic ulcers and gangrenous extremities (feet, fingers etc.) clear up in a matter of weeks. Many individuals were assured that, because of rampant gangrene, limb amputation was the only viable option. Most of these have had not only the smug satisfaction of seeing their doctors proved wrong but of also experiencing the thrill of seeing their extremities completely heal and the gangrene disappear—albeit, some areas of dead tissue having to be trimmed away surgically later. Doctors practicing this therapy the world over have maintained countless files and clinical records to prove, beyond any doubt, that serious cases of arterial disease have been successfully reversed. There are recorded cases of many people being literally at death's door, with advanced chronic diseases caused by blocked arteries, when they commenced chelation. Months and even weeks later, all of them experienced varying degrees of remarkable improvement. Millions of patients have undergone the therapy, probably almost as many as those who have undergone bypass surgery! In addition, several research studies have been published with results which statistically prove that blood flow improves following chelation therapy. Even without blood flow studies, if leg pains are relieved and angina becomes less bothersome or if physical endurance and mental acuity improve, such benefits would more than justify chelation therapy. An improved quality of life and the relief from bothersome symptoms are by far the most important beneficial aspects of EDTA.

What kind of costs are involved? In India the cost of a chelation treatment is 3,000 rupees (approximately $60 US) so a course of 30 chelation therapy treatments would cost 90,000 rupees or approx. $1,800. Costs for treatment in the United States are greater, $100 to $150 per treatment, so a complete course of therapy could run somewhere between

$3,000 and $4500. The costs of the necessary clinical, laboratory and other associated or essential medical tests are not included in these figures. However, this is still considerably less than one shells out for bypass surgery, while it should not be forgotten that ***chelation involves absolutely*** **NO** ***risks*** or the fact that ***surgery does*** **NOT** ***cure the problem anyway.***

What about bypass surgery? Coronary artery bypass surgery has never been proven, even by properly controlled studies, to offer any sort of advantage over non-surgical treatments, besides relieving pain which cannot be controlled with medicine in a minority of patients. This popularly-prescribed procedure involves the blocked portions of the major coronary arteries of the heart being bypassed with grafts from the leg or under-arm veins. It has even been suggested that the relief from pain, following the surgery, is actually the result of cutting the nerve fibres which carry pain impulses from the heart and which also stimulate spasm of coronary arteries. ***It is not possible to perform this type of surgery without interrupting these nerves.*** In addition, arteriograms, which are done to x-ray the arteries prior to surgery, utilize a chemical dye which can cause arterial spasm. In any case, it is difficult to determine from x-rays just how much of the arterial blockage is permanent or what percentage is actually reversible spasm. Indeed, the most recent research suggests that many of the more than 200,000 bypasses performed in all major countries each year for the relief of pain and other symptoms that have resulted due to clogged or blocked arteries, ***are really not necessary***.

A good case against rushing into committing yourself to bypass surgery are the findings of a ten-year, $24 million, study conducted by the US National Institutes of Health (NIH) which compared post-operative survival rates of 'bypassed' patients with a matched group of equally diseased persons who were treated non-surgically. This study proved conclusively that ***there were no all-pervading advantages*** for the majority who had been operated upon when compared with those who received non-surgical therapy. It is important to note that the non-surgical therapies reported in this study did not include either chelation or the new calcium blocker drugs or that only half of the surgery patients had received beta blocker drugs. Although there are studies which show that patients with left main coronary artery blockage live longer after surgery, these studies were carried out much before calcium blocker and the newer beta blocker medicines, which have been scientifically proven to afford protection against heart attack, were made available.

Surgery may even have come out a clear second best if all the presently available non-surgical treatments, including chelation, had been compared with the bypass procedure. When all these findings were statistically analyzed, it was concluded that subjecting a person to surgery ***did not***—in any manner—improve the chances of his being able to live longer, live healthier, live better, or be able to enjoy life more. Most importantly, cardiovascular surgery does nothing to arrest or reverse the underlying diseases, which exist in varying degrees throughout the body. It is, at best, a piecemeal 'cure' for a system-wide problem. Bypassing a restricted portion of the body's blood vessels can have little or no lasting benefits since the same degenerating condition, which caused the most extreme blockage at one or two sites in the first place, must necessarily already be on the rampage everywhere else throughout the circulatory network.

One particular aspect that has been well screened from the general public is the established and documented fact that many, after the first bypass, have had to invariably return sometime later for a second operation! Either it is the blood vessels that weren't originally bypassed, becoming clogged and needing to be cleared, or the transplanted vessels used in the first graft, filling up with new plaque. At other times, it is either the transplants that malfunction or they turn out to be too small for the intended job. As a matter of interest, other studies have shown that by about seven years after surgery, the grafted vessels in 40 percent of the patients had closed completely and in the remaining 60 percent, more than half had developed further coronary narrowing. But once you've attained that bypass 'degree', your chances of making it to 'Ph.D. level' go up by about five percent a year! Five years after the first operation, some specialists estimate, your chances of needing a second could be as high as 30 to 40 percent and there are no guarantees against the possibilities of a third—or even a fourth—operation!

And we've saved the best for last! ***Approximately 3 percent of all bypass patients die as a direct result of the procedure***—even more, if some were already severely ill at the time of surgery. A much higher percentage suffers serious forms of various ***other*** complications, even if they manage to survive the surgery! Chelation patients, on the other hand, are frequently able to return to work and to resume sports and all other normal activities, without any need for surgery. If they maintain a proper diet, exercise regularly, continue to take the prescribed programme of nutritional supplements and have chelation *maintenance* treatments periodically (monthly, more or less, depending on the severity of the underlying medical diagnosis), they are usually able go on for many more years without suffering heart attacks, strokes, senility or gangrenous extremities or even worry about becoming prospective candidates for such attacks. If you have been told, like most people avid for additional information about chelation therapy, that you have advanced arterial disease and that vascular surgery is the inescapable option, ***it is essential that you first understand the nature of the malady and its possible consequences***. Thereafter, 'scour the market' for all possible information on the treatment alternatives available. This will help you to make an intelligent decision on the appropriate option and the required course of action. Even if chelation, or some other non-surgical therapy, somehow fails—***which is rarely the case***—bypass surgery will always be there as a last resort!

Why can't chelation be administered orally in pill form, instead of by intravenous infusion? Chelation therapy is gaining recognition so rapidly that an effective oral **chelator,** which provides similar benefits to intravenous EDTA, will definitely see the light of day very soon. There are many orally administered nutritional substances that are known to have chelating properties but none of these can duplicate the spectrum of activity of EDTA. Many nutrients, such as vitamin C and the amino acids cysteine and aspartic, have the weak ability of chelating metals besides also affording protection against free radical damage as antioxidants. Claims are being increasingly made that, in patients with atherosclerosis, successful use has been made of nutritional supplements containing weak chelators. There is nothing new about these products, which are mostly vitamins and minerals being aggressively marketed with glowing testimonials and deceptive marketing techniques. The benefits from orally ingesting any of these substances has have never even come close to the much more dramatic results seen with EDTA. Recently some nutritional supplements, which contain EDTA, have been ALLEGED to be effective as *oral chelation therapy*. The problem here is that it is *not possible to absorb more than 5 percent or less of EDTA by mouth*. The remainder ineffectively passes out of the body in the stool. If an effective amount of EDTA is to be absorbed orally, there is the need for it to be ingested every day which creates another problem. If taken on a daily basis, oral EDTA will bind to essential nutrients in the digestive tract and block their absorption, causing deficiencies. When given intravenously, it is absorbed 100 percent. Additional nutritional supplementation, taken on a daily basis, will more than compensate for any losses caused by the intravenous infusion.

What are the other diseases that may be tackled by chelation? Because the ageing process itself correlates with ongoing free radical damage, it is no surprise that a large variety of disease symptoms have been reported to improve following chelation therapy, even those symptoms that are not directly caused by circulatory disease. While there is no scientific evidence of chelation being a cure for these diseases, the symptoms of **arthritis, Alzheimer's, Parkinson's, psoriasis, high blood pressure,** and **scleroderma** have all been reported to improve to varying degrees. There is no better treatment for scleroderma; vision has been restored in macular (retina visual acuity) degeneration; patients generally feel younger and more energetic following the therapy—even when it is being taken purely as a preventive measure. In fact, chelation therapy is probably more effective for prevention than it is for eliminating an established disease! A recently published Swiss article from the University of Zurich reports on an 18-year follow-up of a group of 56 chelation therapy patients. When comparing the death rate from cancer between the control group, whose members ***did not*** receive chelation therapy, and the group whose members ***did*** receive EDTA, the authors discovered that the latter group had a *90% reduction of cancer-related deaths*. Epidemiologists from Zurich University reviewed this data and found no faults, either with the reported facts or with the conclusions. There is no evidence that chelation therapy is of benefit in the treatment of advanced cancer, once such a diagnosis has been made, but there is a large body of scientific research indicating that free radical damage to DNA is an important factor in cancer onset. ***Chelation definitely blocks free radical damage.***

Why haven't I heard of chelation before? If EDTA chelation therapy is safe and effective, as indicated by the many published studies and by the experience of hundreds of patients and doctors, why haven't *you* heard more about it? That is a good question! For the moment, suffice it to say that all these new-age therapies are relatively recent arrivals on the Indian medical scene and, until just two or three years ago, only a handful of people knew about their existence. Others received second-hand information that these practices were catching on. Even today, there are a host of medical specialists who haven't heard of the treatments and would probably be reluctant to prescribe them even if they had. The reason for this is that, like many other aspects of our lives, a considerable amount of politics—in this case medical politics—is involved. Orthodox medicine organisations, aided and abetted by the pharmaceutical industry, are politically powerful and have entirely selfish interests. In the western world (especially the United States) and some other advanced countries, such collusive groups have consistently attempted to suppress chelation therapy, perhaps because of a large vested interest in coronary-related health care. The cost of all medical care for victims of heart disease in the United States, including coronary bypass surgery and prescription drugs, exceeds **$40 billion per year** – a pie in the sky that only an idiot would forswear! Obviously, many hospitals and physicians would face serious financial losses, besides trying to find other outlets for their services, *if a far cheaper **and safer** procedure were to become universally popular.* Physicians who remain skeptical about chelation therapy are mainly those who have never tried it and even blankly refuse to do so.

Luckily, we in India may never face such blatant bias against a proven, life-saving therapy. This is probably because, despite our technological advancement and gradual dependence on any and everything newfangled, we have historically been fairly deeply rooted in traditional forms of medicine and healing. From time immemorial we have been very mentally receptive to 'anything' that benefits human health, especially if that 'anything' is invariably and inexorably linked to 'Grandma's remedies'. It is a known fact that a large percentage of our orthodox doctors not only advocate traditional forms of therapy and medicine but many also actively practice them as well. However, since all the laws governing orthodox practices and medicines have actually emanated from the West, and ***they*** being the ones suppressing the spread of chelation, it stands to reason that our home-bound allopathic practitioners have probably not only been denied the required clinical information but may have also been persuaded to ignore chelation. A grossly one-sided and totally biased opinion may have been used as a means of ensuring that chelation remains but a question mark in India. Most doctors are either completely uninformed about the research that has been done to document the safety and effectiveness of the therapy, or they are committed by training to other therapeutic methods, such as vascular surgery and related procedures. Many physicians have merely accepted the criticisms, which are basically of an editorial nature stemming from such sources, without digging into the true facts for themselves. The bypass industry has certainly been extremely well marketed—both to the medical profession and the general public!

What else is involved in a complete course of chelation? Your lifestyle counts. Chelation is only part of the curative process. Improved nutrition and improved lifestyle are absolutely imperative for the lasting benefits from chelation treatments. Chelation is not in, and of, itself a 'cure-all'. It merely reduces abnormal free radical activity, allowing normal healing and the control mechanisms to activate and repair free radical damage. After that, health is restored with the help of applied clinical nutrition, antioxidant supplementation and lifestyle corrections. Chelation certainly instigates all of these factors. It is also compatible with other forms of therapy, ***including bypass surgery!*** In addition to receiving the necessary number of treatments, patients seeking long-term benefits, should correct their dietary and lifestyle routines, regularly take nutritional supplements, be physically active and eliminate destructive habits such as tobacco and excessive alcohol consumption.

It is important to totally eliminate all forms of tobacco use, (cigarettes, pipes, cigars, snuff or chewing tobacco), as it is a consistent observation that the continued use of tobacco, following chelation, lessens the improvement, ***and for far shorter periods*** when compared with non-users. Relatively healthy adults are often able to tolerate the moderate use of alcoholic beverages without generating more free radicals than they can detoxify. Anyone who imbibes more than just the occasional drink is inviting the risk of harmful free radical activity and damage. Victims of chronic degenerative diseases should not only minimize the consumption of alcohol, but give it up altogether, unless it is therapeutically prescribed.

Finally, sustained physical exercise is very helpful. Even a brisk 30 to 45-minute walk every morning (and every evening if possible) will help to maintain the health benefits and improved circulation that have resulted from chelation therapy. **Lactate,** which is a ***natural chelator produced within the body***, normally builds up in tissues during sustained exercise.

Is chelation the therapy for me? Only you can make that decision! **AND** the chances are that your doctor ***will not*** help you to decide! Patients who opt for chelation therapy, often do so against the advice of their personal physicians or cardiologists. Many may have already been advised to undergo vascular surgery. There are a host of people who only heard about chelation therapy after they were hospitalized, or when a friend or relative had urged a look into possible alternative, non-invasive therapies before committing themselves to surgery. It is amazing but true that the majority of first-time chelation patients have invariably ended up for a 'shot' at this therapy on the urging of someone who had already successfully undergone the therapy. So good news *does* spread! True?

What is the procedure - in a nutshell? After making you comfortable, a trained technician will insert a tiny needle, as painlessly as possible, either into a vein on the back of your hand or in the forearm. This will then be taped to prevent it from inadvertently slipping out if jogged. The needle is attached, via tubing, to the EDTA infusion bottle. The drip rate is adjusted as per individual indications, however, the average infusion is about one drop per second. You will be carefully observed during the course of the sitting to ascertain that the infusion is being well tolerated. After every third or fourth treatment, samples of both blood and urine will be taken to ascertain and ensure that the kidneys are functioning adequately enough so that EDTA is safely excreted. At the slightest hint of possible kidney dysfunction, the infusion will be stopped or discontinued until such time that the kidneys resume normal working order. This continuous monitoring is absolutely essential for elderly patients whose kidneys, in the normal course of ageing, usually function at reduced levels. If proper monitoring is carefully exercised, kidney overload is easily prevented. In most cases, kidney function actually improves markedly after a course of chelation. The infusion liquid is normally a solution of about 500 cc, containing EDTA and other additives consisting of various vitamins and minerals. The exact amounts are calculated as per the dictates of the individual's condition. The B-complex vitamins and vitamin C, by itself a weak chelating agent that strengthens EDTA's ability to remove lead and other toxic metals, are common additives as is magnesium, sodium bicarbonate etc. A tiny amount of anti-coagulant prevents clotting at the injection site. In case some pain or discomfort is experienced during the infusion, a drop of local anesthetic will be added. The treatment is generally accomplished without any discomfort, though one should essentially remain comparatively immobile to prevent any inadvertent disruption of the infusion. However, different people tend to feel differently in different situations. The restless or easily bored types should bring along something that will keep them occupied, like some reading material or maybe some important office work that would keep them busy for three or four hours. Otherwise one can pass the time chatting with other patients or watching TV. Figure 7 shows a typical chelation session. The beneficial effects of EDTA become exceedingly evident after the first few sessions and patients thereafter usually tend to look forward to successive treatments—especially when they find a steady progress in the overall improvement of their general conditions. This generally enhances with each succeeding therapy session. All patients are required to sign a treatment consent form before undertaking chelation therapy.

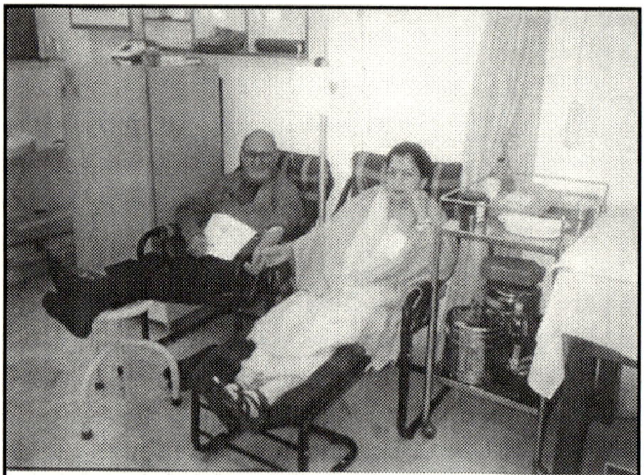

Figure 7 – Receiving Chelation Therapy

What happens if treatment is halted or a person decides to quit? Therapy can be stopped at any juncture with total retention of all the benefits of every session completed up to that point. The benefits normally accrue in direct proportion to the number of treatments received. ***There are no reversal of effects or worsening of symptoms just because the therapy is stopped.*** Once started, the choice of continued periodic chelation is entirely up to the individual. One is also free to change doctors or locations if so desired. A little variation in the treatment protocol will not produce any significant differences. Once recommenced, all the benefits start to add up again with each new treatment—unless, of course, one has delayed far too long between sessions.

How am I to trust the chelating doctor? Can medical competence be judged by a layman? This can only be accomplished if one is bold and forthright. Remember that it is ***your*** personal physical well-being that's at stake here. It will be very difficult for a lay person to judge or make guesses regarding medical competence; however, as with many other medical procedures, ***that shouldn't prevent you from satisfying yourself in regard to the knowledge, experience and credentials of the doctor you have chosen.*** Ask, and continue asking, whatever questions you want till you are completely satisfied. Your physician is duty-bound to answer every query. To start with, you may like to see written proof of his qualifications and training or some records of previous treatments. If you have come this far in the reading, there is sufficient material herein for you to refer to and frame the questions accordingly. Don't be bashful, there are far too many risks involved if you wind up with an incompetent physician. Most importantly, please refrain from signing any treatment consent forms till you have absolutely **NO** doubts that you ***are*** in safe hands. As a means of making sure, some of the telltale omissions that may grab your attention are:

- Failure to ask for your complete medical history including pertinent hospital records, recent x-ray reports, EKGs, arteriograms or other lab investigations.
- Failure to carry out a thorough hands-on physical examination which must include detailed questions regarding living routines and habits.
- Failure to prescribe a pre-treatment comprehensive series of blood and urine tests and some form of non-invasive vascular testing.

And, after commencement of chelation sessions:

- Failure of the doctor or his clinical staff to periodically monitor kidney function through urine and blood tests during treatment.
- Failure of the physician, or any of his staff, to supply asked for information about tests, or to answer questions regarding procedures, symptoms or progress.
- Non-provision of personalized, individual attention. Any *rush jobs* noticed— chelation infusions being completed in one or two hours—are an indication that the well-being of patients at ***this*** clinic is a non-existent item on the agenda.

No treatment is better than the doctor administering it! Unfortunately, just as incompetents and opportunists exist in other fields of medicine, there also are many chelation quacks floating around quite freely, and with a complete set of dubious distinctions in tow! To ensure personal protection and safety, it is expedient that every prospective patient necessarily also be an ***informed*** patient. If your prospective doctor is given to 'hemming and hawing' when questioned—get out of there as fast as you can!

The Proof of the Pudding—!

LCJ is a pleasant 65-year-old who came to us in January 2000 with complaints of angina and breathlessness when walking. His pulse varied between 80 and 96 and he exhibited all the symptoms of hypertension as well. Another unusual complaint was that he could not wear leather footwear because they hurt the soles of his feet. His medical record revealed that in 1987, he'd had an angina attack for which various medicines were prescribed and after a battery of clinical tests he was advised to have an angiography—***which he never had done***. He thereafter went on a strict, controlled diet and maintained his physical health fairly well. As an added precaution, every year he religiously had a thorough physical examination with all the attendant clinical tests wherein the **stress echo test** *was always found to be positive*. Since his medical records were quite dated, we put LCJ through the entire spectrum of blood tests, stress echo, ultrasound of the abdomen, cartiography etc., so that we had the latest picture on what ailed him. The cartiography showed that he suffered from triple vessel disease (TVD).

On Feb 2, 2000, we commenced chelation therapy with EDTA. After the fourth sitting LCJ reported that his soles no longer hurt and that he was once again comfortable wearing leather footwear. He was encouraged to start taking short walks early in the morning and by the tenth treatment session was not only able to walk for 40 minutes without any breathlessness but could also easily climb stairs unaided. Throughout his treatment regimen a peculiarity often noticed was that his pulse rate would register 90 to 96 before the start of therapy but would come down to 72 at the end of treatment. On July 9, 2000, he had completed 20 therapy sessions and treatment was discontinued because he never felt better in his life and was in excellent physical condition. On August 27, 2000, LCJ undertook an Executive Health Check-up Programme, which entailed the entire gamut of tests, at one of the larger hospitals at his own behest. Interestingly, the stress echo test proved negative for exercise-induced wall motion abnormality. There was also a good increase in global LV systolic function and LVEF had increased from 66% to 78%. Following these results, his cardiologist stopped all the medications that had sustained him till now.

Four months later, we received a sudden call from LCJ's home that he was extremely unwell. On reaching the house, we were informed that LCJ had felt somewhat ill in the morning and had taken a sorbitrate following which he had blacked out. When LCJ's pulse was found to be 140, he was immediately moved to a hospital for medical management. He was put on a course of metaprolol and his condition stabilized by the next morning. The attending physician, who was also a cardiologist and an expert on fibrillation, suggested that LCJ undergo an angiography as it was suspected that a blood clot could have been the cause of the problem. On our advice, this time LCJ agreed to the procedure which was carried out that very evening. Subsequently, we received the elating news that all of LCJ's arteries were completely clean and their condition was akin to that of a 20-year-old. Unfortunately, his pulse rate problem had persisted and had probably blossomed because he was completely off all medication. However, that can't take away the kudos for EDTA which had performed its job with aplomb and returned LCJ's health to him on a silver platter.

What's the Last Word, Then?

To sum up, you now have to contend with smog-filled skies and fat-filled foods—how best do you fend off external and internal pollution—what must you do to retain the health benefits of a completed course of EDTA infusions? On the brighter side, the chances are that you're breathing easier, walking further, have much more energy, and you're also looking younger and feeling sexier than you have in years! On the gloomier aspect, you're far from being home-free, because not even EDTA can provide that cast-iron guarantee. Being chelated was probably the easy part but the real challenge comes in remaining detoxified, preventing new plaque accumulations from clogging your arteries, maintaining an efficient immune system and protecting your cells from further free radical damage. This will certainly take some doing on your part and there's much more involved than just coming in for those regular *maintenance* sessions.

You probably went into a tizzy when you learned that some serious ailment was about to disrupt your life and we doubt that you welcomed whatever the health problems were that precipitated action regarding chelation. But if you have opted for this therapy, consider yourself fortunate that you took advantage of the timely warning. Not too long ago, any diagnosis of serious degenerative disease was akin to getting the death sentence! At best, the 'accused' was condemned to suffer the endless rounds of doctors and drugs. It was the same for those who had reached '*that*' particular age; the only thing occupying their minds was that they had passed fifty and everything was downhill from there on since nothing could be done to slow the ravages of time. But now that you've *'met'* chelation, you know better. The odds of your living longer and in a healthier manner improve when you have accurate information and act upon it. ***The more you know, the better your opportunities to make the correct health-promoting lifestyle choices.***

CHAPTER 6

DIMETHYL SULFOXIDE (DMSO) – A BOUNTIFUL GIFT FROM NATURE

Ouch! Stroked 'Out' in One Stroke!

In July 2000, AS underwent coronary artery bypass surgery. This commonly administered surgical procedure appeared to be routine and uneventful for the patient. Approximately 200,000 such open heart surgeries are routinely performed in India every year. After a week's post-operative hospitalization, AS returned home to continue with his recovery. On a Saturday evening about a month later, he went out to dine at a popular 5-star restaurant with family and friends. It was during dinner that sudden complications set in for the still-recuperating patient with AS evincing all the signs of undergoing what is commonly known as a **stroke**. Medically called a **cerebrovascular accident** (CVA), this occurs from an **embolus** (blood clot) or bleeding within the brain which causes a lack of oxygen in the tissues which are normally supplied by these same leaking blood vessels. The immediate after-effects of a stroke depend on the location and extent of damage, and in AS they became evident as an obvious speech defect (**aphasia**), an inability to understand language being directed to him, and a feeling of extreme weakness and paralysis affecting the left side of his body. As diagnosed later by his cardiologist, these various manifestations of stroke had struck because one or more emboli may have flicked off his coronary artery. "I felt many sensations in that restaurant: disorientation, my jaw dropping, drooling from the corners of my mouth, paralysis creeping over my extremities, and worse," AS recalled. "Dr. Mirchandani was also present at that dinner and witnessed what transpired. He got us back home and immediately set me up on an intravenous drip using 8 cc of DMSO with other IV multiple vitamins and minerals mixed in 500 cc of 5% dextrose solution. Dr. Mirchandani maintained me on this mainly DMSO infusion for over a three-hour period. Two hours into the infusion, all signs and symptoms of the stroke had totally subsided. On the following Sunday morning I was given a second infusion of DMSO and felt my condition improving further."

"On Monday morning," he continued, "I went to the hospital to have an MRI (**magnetic resonance imaging**) done. After the test, though the specialists concerned were all in agreement that I had indeed been struck by a CVA, they were completely unable to locate the site of the stroke. It just didn't show up anywhere and my blood vessels were as clear as if nothing had happened. When the hospital's physician (internist), ***and*** my cardiologist, learned of the IV procedures that I'd gone through, I was told, 'You can't do that! Nowhere in regular medicine are physicians trained in the IV application of DMSO. We don't believe in it.' The intern made that silly statement despite my standing before him as living proof that, far from doing me harm, DMSO had worked immediate wonders as a rehabilitative agent for the stroke. Despite my cardiologist's vehement protests, I've since had seventeen more DMSO treatments and have physically improved so dramatically that, at this moment in time, there are absolutely no residual symptoms or signs of the stroke remaining."

Now Where Did This New 'Thing' Come From?

Dimethyl sulfoxide (DMSO) was first synthesized in 1866 by Russian scientist Alexander Saytzeff, who noted that the substance was colourless, smelt like garlic, was oily to the touch, resembled mineral oil when poured and tasted like clams or oysters. It combined with almost any chemical and was an excellent solvent proving extremely useful as a de-greaser, paint thinner and anti-freeze. It was only after World War II that chemists started to show an active interest in the substance, and in 1959, British chemists demonstrated that DMSO protected red blood cells and other tissues against freezing conditions. It is a reagent that can speed up some chemical reactions a billion fold. Basically the substance is a simple by-product of the paper-making process. It can be derived from **lignin**, the cement substance of trees, and can also be synthesized from coal, petroleum or other organic substances.

What amazed scientists and chemists alike was the liquid's tremendous capacity to dissolve substances. DMSO's unique capability to penetrate living tissues ***without*** causing significant damage is related to its relatively compact structure and polar nature as well as its capacity to accept hydrogen bonds. This combination of properties allows it to associate with water, proteins, carbohydrates, nucleic acid, ionic substances and other constituents of living systems. In the understanding of the possible functions of DMSO in biological systems, of premier importance is its ability to replace some of the water molecules associated with the cellular constituents. But what really caused a scientific 'buzz' was its drying effect. Since wetness or moisture generally tends to instigate infection, the dryness of a therapeutic agent makes it very valuable in the treatment of burns. Laboratory tests proved that DMSO easily passed through skin and mucous membranes with the added advantage of *being able to carry along with it* other substances as well during its passage. For example, if either a local anesthetic or penicillin is dissolved in DMSO they can be ***carried through the skin and into all organs of the body without using a needle***! DMSO was first used as a pharmacologic agent in 1962 and gained fame as a chemical that could relieve inflammation and pain in many conditions, ***including those that had remained untreatable by all other methods.*** Early studies had revealed that it could relieve pain, reduce swelling, hamper bacterial growth, soften scar tissue, improve blood supply, was an effective muscle relaxant, could act as a diuretic and could also enhance the effectiveness of other pharmacologic agents. In addition, the pain of sprains, burns, strains, arthritis—***even those of broken bones***— was completely eliminated. DMSO kills viruses and fungi and works marvelously against cataracts, sports injuries, scleroderma, myasthenia gravis and tuberculosis. In persons with Down's syndrome, the mental retardation has been considerably lessened. It has been popularly prescribed by vets for arthritic conditions and injuries in animals.

The Working Environment

One fascinating property of DMSO is its ability to associate with molecules that have a thick layer of hydrogen ions and with neutral molecules as well as ionic species. This ability is what makes it such an excellent solvent that can easily penetrate through both organic and some inorganic materials. It bonds strongly with the hydroxyl groups. The significance of DMSO as a scavenger of hydroxyl radicals, which are the main culprits that break down the synovial fluid as well as the cartilage of the joints, is that ***these are the dominant chemical ions in arthritis*** and DMSO is one of the few known substances that successfully detoxifies these radicals. It forms a dimethyl sulfone plus water with the hydroxyl ion which is then readily excreted out of the body. The neutralization of this highly toxic *free radical* is what results in the reduction of inflammation and the lessening of arthritic pain. This is the primary mechanism that permits DMSO to work so effectively against arthritis.

Another extremely important aspect of DMSO is that it works in a similar therapeutic manner against the pathological biochemical changes in cancer, atherosclerosis and any other set of circumstances where the generation of free radicals is all-pervading. As you now know, free radical pathology is an intricate part of every metabolic dysfunction that can be thought of. Cancer manufactures enormous amounts of hydrogen peroxide, lipid peroxidation takes place and more hydroxyl radicals develop resulting in the onset of cellular damage and degenerative disease. DMSO offsets these effects and brings the body back to a more normal state, which is probably part of the reason why it has proved to be so valuable in the treatment of cancer. In experiments carried out at Mount Sinai Hospital in New York City, cancerous cells have been turned into normal, harmless ones by treating the diseased cells with DMSO solutions.

Periodontists (the tooth 'fairies') have cleared up gum diseases and reduced tooth decay by painting the affected areas with DMSO solutions. Russian doctors have successfully treated generalized periodontitis with a mixture of indomethacin (an anti-inflammatory, anti-fever and painkilling drug) suspended in DMSO. Periodontitis is a disease of the structures supporting the teeth (gums, periodontal membrane and alveolar bone). Bacteria, acting on food debris imbedded between the teeth and gums, cause the formation of plaque or tartar which tends to accumulate in the **gingival crevices** which are spaces between the gums and the surface of the teeth. These spaces, which eventually become abnormally enlarged forming **gingival pockets,** are an early indication of periodontal disease. In chronic cases, the gums become highly inflamed, swollen and they bleed easily. The accumulation of tartar causes ulceration and further bleeding. If this is not treated in time, the plaque will then spread to, infect and then destroy the periodontal membrane and alveolar bone which results in the teeth eventually loosening and falling out. This disease is the major cause of tooth loss in the middle aged and elderly and may be brought on by either poor dental hygiene or by badly fitted dentures and badly made crowns and fillings.

Podiatrists (foot doctors) have successfully used DMSO in the treatment of corns, calluses, ingrown toenails, bunions, hammertoes, heel spurs and even the inflammation of the big toes due to gout. Gout pain usually disappears within seven days of the first treatment. The most amazing versatility of DMSO remains the speed at which it is easily absorbed into the body. If you paint a coin-sized amount anywhere on your upper body, it will penetrate the skin and travel through the bloodstream so fast that within just thirty seconds you'll taste it on the tip of your tongue!

Does It have Any Other Properties?

Definitely so! The most important one is its ability to function as a 'drug transporter'. Most of the drugs that are dissolved in DMSO not only retain their therapeutic activity but also their specific individual properties over a long period of time. DMSO has been seen to enhance and multiply the actions of drugs that are dissolved in it thus permitting smaller drug doses to be administered without any reduction in their therapeutic efficacy. In addition, it has been seen that the bad side-effects of most drugs have been greatly diminished after they have been mixed with it. In organ banks around the world, tissues and organs stored as transplants or grafts, are generally preserved in DMSO solutions. As a penetrating carrier of other drugs, it has no equal. It carries the necessary pharmaceuticals, for the required therapeutic effects, to any part of the body passing easily through cellular membranes and tissues without causing any harm. It is able to penetrate endothelial coatings of the arterial walls, meninges of the brain, healthy skin, mucous membranes and other tissues. An intramuscular or intravenous injection of DMSO passes without problem into the fluid of the head and spine. When it is injected within a horny case, like that which surrounds a muscle or nerve, it rapidly appears in the bloodstream. The response of the central nervous system to DMSO is very different from those of its responses to other drugs because DMSO passes the blood-brain barrier, easily penetrating it and passing out again. This blood-brain barrier is normally impenetrable by other drugs. However if these other drugs happen to be molecularly mixed with this solvent, they can also now surmount this obstacle with its help. DMSO stimulates the basic chemical processes at the nerve cell level in the central nervous system. It allows the transport of amino acids to the brain which energizes both the functional activity of the neurons as well as the brain itself. It is this functional stimulation that permits corrections of the many neurological syndromes characterized by mental deficiency, slower brain activity, loss of memory, depression and anguish.

So Many Properties Mean Plenty of Other Actions!

Very true! DMSO has a wide range of pharmacological actions including membrane penetration and anti-inflammatory plus local analgesic effects. It inhibits the growth of bacteria, promotes urine excretion, stems cholinesterase secretion, changes the action of drugs that are mixed with it, is a collagen solvent, provides a specific or nonspecific increase of immunity and produces local vasodilation. DMSO aids in permitting incisions without pain as it effectively carries local anesthetics into deeper layers of the skin and body. It displays far more power when it is integrated with other drugs than it would as the sole medication. When administered alone, DMSO had limited effects on patients treated for cellulitis (inflammation of cellular or connective tissue). However, when combined with an antibiotic, although DMSO already possesses antibiotic properties, the results were rather dramatic. DMSO can convert bacteria which are resistant to a given antibiotic into becoming sensitive to that same antibiotic! This trait of being able to re-sensitize bacteria could possibly see the resurgence of some groups of antibiotics that are currently obsolete. A point to note is that though DMSO readily crosses most membranes of the body without destroying their integrity, it does not rapidly or readily penetrate the nails, the hair or the enamel of the teeth.

Summary of Its General Medical Uses

DMSO has at least forty major therapeutic properties which act beneficially in the process of healing traumas to the body and brain. Here are sixteen of the more common ones:

1. It blocks the transfer of pain messages, from the site of injury to the brain, by interrupting conduction in the small non-myelinated nerve fibres. Thus, it is one of the finest topically-applied pain killers known to man.
2. It is potently anti-inflammatory and marvelous as an anti-arthritic remedy.
3. Since it is **bacteriostatic**, **fungistatic** and **virostatic** it inhibits the growth of pathogenic micro-organisms.
4. It transports numerous pharmaceuticals across body membranes, thereby administering them topically without resorting to injections.
5. It reduces the incidence of **platelet thrombi** in blood vessels thus preventing or reducing the formation of blood clots.

6. It has a specific effect on cardiac contractility by holding back calcium from muscle fibre absorption, thus reducing the workload of the heart.
7. It acts as a tranquilizer, producing brain sedation, even when simply rubbed into the skin.
8. It enhances the actions of anti-fungal and antibacterial agents when combined with them for internal or external application.
9. It is a **vasodilator**, probably related to **histamine** release in the cells and to **prostaglandin** inhibition.
10. It stops the release of cholinesterase, an enzyme that causes the breakdown of acetylcholine, which is one of the B complex vitamin forms necessary for healthy neuronal functions.
11. It tends to soften **collagen** as a means of preventing or reducing scars and **keloids** by its peculiar cross-linking effects
12. It scavenges the hydroxyl free radical so as to reduce tissue deterioration in ageing.
13. It stimulates various types of immunity, especially the blood's natural killer cells that are derived from **lymphocytes.**
14. It is a potent **diuretic**, particularly when administered intravenously.
15. It brings about **interferon** formation in the organism as an express means of boosting immunological response.
16. It stimulates the healing of wounds of all types, both external and internal.

This drug has the widest range and greatest number of therapeutic actions ever shown for any single chemical. It also has the ability to grow cumulative in its effects. Since it attacks the disease itself rather than just the symptoms, it has been observed that lesser amounts of DMSO are needed to achieve results with the passage of time. This is truly the unique and different quality of this chemical, since most other drugs generally require increasingly heavier dosages over time to maintain therapeutic levels. There is absolutely no doubt that it has proved to have interesting and valuable biological properties.

So How Does It Tackle Various Maladies?

At 67 years of age, CLJ was in a truly pathetic physical condition with a veritable plethora of complaints. Though four years earlier he had undergone bypass surgery he continued to experience angina pains as well as a choking sensation in the throat during exertion. He walked unsteadily and was unable to bend his knees due to the onset of **osteoarthritis**. He also had hypertension and the remnants of right upper limb paresis. The throat choking symptoms had first surfaced in 1993, along with angina attacks, for which various medications had been prescribed. Because of the angina attacks, an angiography was performed which revealed that he had 40 percent ***stenosis of the left main artery***, ***triple vessel disease,*** as well as ***LV dysfunction***. Bypass surgery was recommended and eventually performed in December, 1995. The procedure went off without a hitch, but during the post-operative recovery period he developed right upper limb paresis for which he was put on physiotherapy. After discharge from the hospital, he strictly followed the prescribed regimen of medication, diet and exercise. However, nothing had been done about the limb paresis at the hospital and it took three whole months of regular physiotherapy for the limb to regain a slight semblance of movement. Four years later the very same symptoms of his initial complaints, ***which should have been eradicated via surgery***, reappeared in full force. Due to the recurring chest discomfort and throat pain, he underwent an investigative stress ECG which proved to be 'abnormal'. On tabulation of the various findings and other examinations, he was advised by his doctors ***to undergo a further angiography for confirmation purposes*** and follow this up with ***a second bypass surgery***. CLJ vehemently refused to undergo any more surgery and insisted that he continue with medications only. His doctors had no alternative but to agree.

He came to us in September 2000, and we commenced initial therapy with chelation treatment which lasted for eleven sittings. This was followed by five infusions of DMSO. Thereafter 25 treatment sessions were given alternating DMSO and chelation. His overall physical improvement was not only truly remarkable but almost unbelievable! He was able to walk again without any angina or fatigue and the throat choking symptoms also disappeared. The osteoarthritis has improved to the extent that currently there are no impediments to free movement. Best of all, the paresis of the right upper limb, which had continued to persist even after physiotherapy, has ***completely vanished!***

Wow! Versatility Unlimited – Right?

It is definitely very much more than that! Since there are several hundred therapeutic uses of DMSO, the following—prepared primarily for the lay person's medical armamentarium—is merely a partial listing of its myriad applications. To effectively highlight the true efficacy of this wonder drug, it has been necessary to elaborate certain details of what happens when various diseases and injuries are treated with DMSO.

CANCER. As an excellent adjunct in metabolic cancer therapy, DMSO enhances chemotherapy by carrying the chemo serum to the exact site in the quickest possible time. The combination normally has an enhanced antiblastic activity (destruction of cancer cells) due to the potentiating and penetrating property of DMSO and its added capability of considerably reducing the toxic side effects of chemotherapy agents employed in prolonged treatments. DMSO and **hematoxylon**, administered intravenously as a combination, have excellent tumor shrinking qualities. This mixture has marvelous therapeutic corrective effects especially for **adenocarcinoma**, **squamous cell carcinoma**, **prostate cancer**, **breast cancer**, **lymphoma** and some other solid tumors. DMSO has been shown to effectively induce anti-tumor activity of **cyclophosphamide** in the treatment of both lung cancer and brain tumors. Patients with **multiple myeloma**—a cancer of the kidney—have also been successfully treated. The DMSO-hematoxylon combination is known to have a great affinity for tumors without causing any ill-effects elsewhere in the body. No matter where the combination is injected, it will eventually follow the bloodstream and unerringly target and destroy the tumor ***without flowing anywhere else in the body.*** In **lymphosarcoma,** the cancer cells are fed by an intestinal material called **protoplasm** which provides the nutrition for the cancer cells to grow. The DMSO-hematoxylon solution removes this nutritional substance causing the cancer cells to starve to death. ***DMSO by itself cannot affect or reduce tumor cells.*** It is the *hematoxylon* (or any other chemotherapeutic agent) that is the actual destroyer. However, without the aid of DMSO's wonderful and remarkable tissue penetration ability, chemo agents would not be able to easily reach the site in concentrated form to more effectively perform the required therapeutic actions.

DMSO has a therapeutic use for skin cancer. Its effects for this treatment are in some respects similar to those of the skin substances called **retinoids**. Topical applications have been used in many open lesions of malignancies with encouraging results and without any toxicity being evident. The *DMSO-hematoxylon* combination removes cancer pain and reduces or eliminates free radical pathology that is a thriving characteristic in most cancer patients. Chemotherapy or radiation therapy generally creates vast amounts of free radicals in cancer patients which manifest themselves as a metallic taste in the mouth, dry mouth, sores at the corners of the lips, nausea, hair loss and other symptoms. DMSO has been seen to relieve and reduce many of these free radical signs and symptoms.

ARTHRITIS. In its entirety, this is basically a degenerative disease that affects the joints and the bones, and also causes musculoskeletal disorders. Though there are many, many variants, each with its own particular problems, symptoms and treatment, the two most common forms of this disease are **osteoarthritis** and **rheumatoid arthritis**. You may also have heard of **gout** and **spondylitis** which are both part of the 'inner circle' of this vicious disease. The onset of the malady is secretive and subtle, invariably going unnoticed until it flares into full maturity. There are various symptoms and signs, quite common to practically all the forms which may point to its possible presence, but these are hard to confirm in the starting stages. The main symptoms of arthritis are joint discomfort and pain. This could swing from occasional stiffness and mild twinges to never-ending severe pain, swollen joints and loss of mobility. In other variants, like rheumatoid arthritis, there would also be fatigue, fever, weight loss, malaise, loss of appetite, anemia and evidence of ruptured tendons. However, the predominant and overriding symptom is pain in varying degrees.

As a simple explanation, consider that every joint in your body comes equipped with its own set of 'shock absorbers'. The first of these is **cartilage** which caps the ends of the bones in a healthy joint. Cartilage is a sort of cushion of smooth, white, resilient but extremely strong material that prevents the bones from grinding against each other. The second 'shock absorber' is the sinewy **connective tissue**, which is moist and flexible, that holds the joint together. **Synovial fluid** lubricates the joint and provides some added cushioning. When these natural shock absorbers deteriorate or become inflamed, preventing the joint from functioning properly and giving rise to pain, you can bet your bottom dollar that arthritis is the culprit. Another piece of bad news is that arthritis gets worse over time. Unlike an infection that the body can fight and get rid of, this is a disease that progressively deforms and ultimately destroys the joints. So, what can you do about arthritis? Until recently, nothing much could be done except for resorting to analgesics to relieve some of the pain or going in for surgery when things got really bad. But now DMSO can step in.

Mrs. K, a qualified doctor, had a chronic arthritis problem for over a decade. The disease was so far gone that she was unable to bend her left knee and was severely restricted for movement as any motion caused the most excruciating pain. With her medical qualifications and connections, she was in a position to obtain the best of medicines, treatment and advice, but to absolutely no avail. The disease just progressed and her condition worsened. When she eventually came to our clinic she was initially given DMSO for simple topical application. The result was dramatic. By simply applying DMSO (in cream/salve form) over the affected area for a few days, the pain disappeared and the knee that she had not been able to bend for years regained movement. Subsequently, she was put on intravenous treatment. A 5% dextrose solution with 2.5 ml to 6 ml of DMSO along with multivitamins and multiminerals was the intravenous protocol prepared for her. After ten therapy sessions she is now completely free of pain and is able to stand and move about for hours on end without any discomfort. Topical applications of DMSO, in a specially developed and highly effective cream form continue as a regular feature.

Arthritis is one of those degenerative diseases listed as "having no known cause and no known cures". Age plays an important role and, generally, the older one is the more likely is he or she to contract the disease. Whereas osteoarthritis normally strikes those who are past mid-life, rheumatoid arthritis commonly shows up before mid-life. Gender is also a governing factor because female victims of arthritis far outnumber males. Wearing high heels has been identified as a cause for osteoarthritis of the knee. This happens because the knee is unnaturally stressed by having to compensate for the changed position of the ankle. Those having a family history of the malady are also more than likely to become arthritic through heredity. Being obese and overweight doesn't help matters either. Extra weight increases stress on the joints making cartilage and connective tissue less able to replenish themselves. People whose jobs entail high or unusual physical demands (athletes, dancers, manual laborers spending long hours in factories or industries) are prime candidates as well. Last but not least, any injury or damage to the cartilage or connective tissue could prevent effective replenishment from reaching them resulting in the injured area becoming arthritic at some later date.

But how does DMSO perform so well in these cases? The exact mechanism is not known. However, since it is a great vasodilator as well as a marvelous solvent capable of penetrating membrane and tissue without causing damage, it is estimated that DMSO helps by relieving pressure and reducing inflammation in the affected area, soothing pain-causing nerves and helping the body in the production of required nutrients and fluids. The nutrients rejuvenate cartilage and connective tissue production/replenishment as well as stimulate synovial fluid lubrication. DMSO effectiveness is enhanced considerably when the therapy is combined with a scientific diet regimen which will ensure that patients receive the nutrients specific to the type of ailment. So, a patient first needs to determine exactly what type of arthritis he or she has. This is necessary because nutrients that work for osteoarthritis aren't useful for rheumatoid arthritis, and nutrients helpful to rheumatoid patients have proved ineffective with osteoarthritis.

Brain and Spinal Cord Injuries. In the same way that DMSO reverses the effects of embolic or hemorrhagic stroke, it is superior to any other treatment for high velocity missile (gunshot) wounds to the brain. It helps in the prevention of paralysis after injuries to the brain or spinal chord. It is necessary to orally administer DMSO by mouth and intravenously as soon as possible after the head wound occurs to reduce brain pressure to normal. This is because, after every injury the grey matter of the brain seems to deteriorate. The healing qualities of DMSO also work quickly to bring injured spinal cord tissues back to normal. The relief comes about by the solution being able to rapidly cross skin and blood-brain barriers without damaging the cells. In addition, it also protects the cells from mechanical damage. In head injuries and strokes, much of the permanent neurological damage is caused by reduced blood flow to the brain which results in an inadequate supply of oxygen and nutrients to brain tissue. DMSO allows a smoother and better blood flow by lessening the stickiness of *platelets*, inhibiting their assemblage and reducing clot formations, conditions that would ordinarily clog arteries and veins. It also stimulates the release of a prostaglandin which increases blood vessel diameter and blocks the release of *another prostaglandin* which tends to constrict vessels and hamper blood flow. **Prostaglandins (PGs)** are a group of cyclical fatty acids that possess varied and strong biological activities that affect cell function in every organ system. They have sedative, tranquilizing and anti-convulsive effects on the central nervous system.

When brain tissue dies because the blood supply has been cut off, **occlusive infarction** of the brain results. Blood carrying supplies of oxygen and nutrients becomes blocked by a blood clot called a **thrombus** or by a plug of clotted blood or foreign materials termed an **embolus**. These brain *emboli* can come from fatty plaques in the large arteries that supply the brain; they may be pieces that break off from *thrombi* in damaged hearts; they could form from *platelets, fibrin* and *cholesterol* that have piled up on the surface of ulcerated plaques in arteriosclerosis; from mural *thrombi* in rheumatic heart disease or following a heart attack; or from blood clots following open heart surgery. If one these emboli blocks the blood supply for more than a few minutes, a small portion of the brain will die and the resulting neurological damage is usually permanent. Another symptom of blood supply blockage to the brain is **acute stroke**, a cerebrovascular disease that is the most common cause of neurologic disability. Stroke manifests itself mostly because of arteriosclerosis, or high blood pressure, or a combination of the two and can strike quite suddenly—the accompanying symptoms and signs being a general indication of the particular area of the brain that has suffered damage. If either the internal carotid artery or the middle cerebral artery gets occluded, a severe set of neurological abnormalities normally result.

The pathological sequence that results from cerebrovascular insufficiency finally ends up with a localized brain area dying from an embolus. DMSO prevents or reverses this sequence if it can get to the site of the injury fast enough. It stimulates PG synthesis which stops, and even reverses, the potential pathological damage synonymous with brain infarction. Through vasodilation of other blood vessels, besides the one blocked by the embolus, DMSO increases the oxygen available to the brain. It also aids in the release of energy that gives sufficient time for the ischemic tissue to achieve a *hemostatic* balance and either reverse or prevent further damage to the central nervous system. Following a head, or spinal cord, injury there is always an increase in blood pressure which could lead to death if left unchecked. DMSO stabilizes elevated blood pressure. It also increases the respiratory pattern; patients start breathing harder and faster, which is a beneficial effect because in many brain injuries respiration tends to become increasingly shallow and could ultimately stop. In the urine output, it has a diuretic effect that almost trebles the body's excretion prowess. The damage from head trauma breaks down many cells causing them to spill their contents into the tissue resulting in water accumulation in and an increase of pressure on the brain. This fluid build up eventually compresses vital centres at the bottom of the brain and may lead to death. DMSO reduces intracranial pressure by picking up this excess water, carrying it to the blood vessels and finally removing it from the brain. In a like manner, an accumulation of blood that compresses tissues in an area critical to survival can cause unalterable brain damage. DMSO has the ability to remove even this non-toxic blood that has gathered at a wrong place. In the area of spinal cord trauma, intravenous DMSO immediately sets about elevating the blood flow in the region preventing the onset of paralysis as well as the spillage of necessary enzymes and other materials into the tissue.

DOWN'S SYNDROME. DMSO has proved very useful in Down's syndrome and mental retardation cases. For brain damaged children, administered as a 50 percent diluted drink given in the equivalent of one half of a gram per kg, it has been seen to speed up the normalization process in affected children. As a penetrant, it carries drugs across the blood-brain barrier which has always been a great problem when the brain is being treated. The many retarded children (normally labeled **trisomic mongoloids**) who were regularly treated with DMSO, not only reverted to having almost normal features and functions but also mentally improved to an extent where evidence of retardation became almost non-existent.

SCIATICA AND LUMBAR DISC PROBLEMS. When 20 to 50 cc of 20 percent DMSO combined with the local anesthetic **xylocaine** are injected for three to five consecutive days intramuscularly into a painful back area, the pain disappears and remains gone for longer intervals between series of injections. A 20 percent strength of pharmaceutical grade DMSO injected into diseased spinal discs rids the patient of nerve pain from the damaged discs.

KELOIDS, SCARS, ACNE AND BURNS. Applied topically and repeatedly, DMSO flattens the raised, nodular, lobulated linear mass of scar tissue in **keloids** and also takes away much of their discolouration. It also reduces the scarring caused by **chronic acne** where fibrous papules develop at the site of hair follicles, normally at the hairline on the nape of the neck. DMSO is unrivaled in the treatment of burns. Not only does it considerably reduce burn pain, but is instrumental in the speedy healing of damaged areas. It reduces, dissolves or prevents the formation of scar tissue—a very important feature in burns as it prevents the contracture of the scar tissue normally left after such injuries.

MUSCULOSKELTAL AND SPORTS INJURIES. Full-strength DMSO painted on sprained ankles, pulled muscles, dislocated joints and even at the site of simple fractures removes pain like magic. It is a highly effective healing tool in the sports arena and is heavily relied upon by trainers and coaches in the treatment of soft tissue injuries where professional athletes are concerned. DMSO spurs the healing of almost any type of traumatic occurrence to the musculoskeletal system. For example, if applied to the back immediately following an injury the healing process is triggered almost immediately preventing any onset of paralysis. Chronic conditions, which may have become acute again after having previously been eliminated, respond rapidly with relieved pain, reduced swelling and improved function. It has also promoted rapid recovery and return to action of cases that required immobilization following fractures. The simple explanation for these dramatic and positive effects is that DMSO somehow revitalizes and returns to their normal state those cells that have been damaged by **osmotic stress.** Surprisingly, when applied topically, DMSO is better tolerated in higher concentrations below the waist rather than above the waist. Yet it is more effective above the waist, having the quickest response for problems involving the face, neck, shoulders, upper limbs and upper back and trunk. It should be used with caution because it can cause side effects such as a burning sensation, itching or redness of skin area. If any such symptoms manifest themselves or discomfort occurs, the solution's concentration should be diluted, preferably with distilled water.

FOOT DISORDERS. The greatest patient successes experienced by podiatrists have been the application of topical DMSO for pain relief, healing and cures in a host of foot disorders. Some of these are explained below:

- **Big toe bunions**. These normally develop due to the deviation of the two adjoining bones at the joint of the big toe and the first metatarsal and occur in two forms, acute and chronic. The acute bunion, or **bursitis**, is the sudden and painful inflammation of a sac that acts as a lubricant between skin and bones. Ill-fitting shoes are a major cause. If not treated properly, bursitis will slowly develop into the chronic type, or **hallux valgus,** which produces an abnormal prominence of the big toe joint. This results in the widening of the foot and brings about a loss of balance besides making standing and walking difficult. This condition could result in an early in life development of foot arthritis.

- **Hammer-toes**. This most painful foot ailment is characterized by contraction of the tendons on top of the foot accompanied by the laxity of the ligaments and the second and third phalanges (the individual bones of each toe) which become angulated. Tight shoes compress the feet and constrict the toe muscles which waste away as a result. Toe motions become puny and weak and the affected digit becomes deformed. The nail can split and grow inward and a hard corn may form. Although any one of the toes can be affected, the second toe is most prone as it is usually the longest.

- **Arthritis of the feet**. This results more often from mechanical strain rather than from infection. This condition is the main contributor of other deformities such as fallen arches, high-arched foot, hammertoes, bunions and heel spurs. This arthritis is also known to be associated with knock-knees, bow legs, flat feet and weak feet. Although arthritis of the feet is not curable at this moment in time, the pain and discomforts can be relieved to a great degree.

- **Hard and soft corns**. Hard corns generally form on the top of the toes and are growths of horny skin. Soft corns form between the toes. They result from friction and pressure that arises from an underlying prominence of bone.. The sharp edge of the toe bone rubs the skin from the inside while ill-fitting shoes may cause friction on the outside. The affected piece of skin eventually dies and builds up in layers creating the corn. Both types are easily distinguishable from the surrounding tissue. Inside a corn is a central **radix**, or eye, of hard grey skin around which is a painful, inflamed and raised ring of skin and flesh that is yellow in colour. Irritation and pain are the normal symptoms.

- **Calluses**. These are thickened masses of skin formed on the load-bearing portions of the feet and are a result of constant friction and pressure. These types of plantar calluses, like the unhealthy one that develops on the ball of the foot, are surrounded by an inflamed red rim and fluid seeping out of the tissues beneath. It is often swollen, hot and painful, and frequently appears as a hard, dry, yellowish horn-like mass with a thick centre that tapers towards the edges. The central portion is pebble-like in appearance and feel. Since the thickened flesh loses its ability to stretch when the feet are flexed, the calluses always give a sort of burning sensation.

- **Plantar warts.** These are benign, tumor-like growths that are well supplied with blood vessels and nerves and grow on the soles of the feet as raised lumps of flesh. They may sometimes be mistaken for corns or calluses because their appearance and symptoms are the same. An injury to the foot may have permitted the entry of a virus that has caused the plantar wart to grow. Walking flattens the warts so that they remain thickened and rough in texture, varying in size from a pinhead to the size of a Rupee and growing either singularly or in clusters. They are extremely painful when pinched. DMSO can only remove the pain and does not cure them.

- **Ingrown toenails**. They occur when the lateral edges of the nail penetrate and cut into the flesh of the toe and can result in severe complications—inflammation, infection, ulceration and gangrene will follow each other progressively—if corrective measures are not initiated immediately.

- **Club nails**. The elderly sometimes have thick, ugly, deformed toenails that indicate either a systemic disease or a chronic injury. Overgrown toenails, that tend to curl under the toes and have become extremely hard, are known as club nails and usually have grotesque shapes. They are both painful and uncomfortable and could provide a nesting place for parasitic fungi. They need to be trimmed and pared in order to have room within shoes, to relieve pressure and to prevent corns from forming on adjacent toes. Besides softening the nails and reducing the sensitivity at their roots so that clipping becomes easy, DMSO also provides a prolonged soothing effect and brings the formerly compressed tissues under the feet back to a more normal state.

- **Fungus toenails**. The most common of nail maladies is fungus-infected toenails caused by parasites such as yeasts, molds and fungi, all of which grow as ringworm. Though the infection may be painless, infected nails are a precursor to athlete's foot and the disease should be treated without delay. DMSO works very effectively for this problem.

- **Athlete's foot**. Medically known as **tinea pedis** (fungus of the foot), fungus by itself does not create this foot disease. The term describes the set of symptoms and not the main cause. Once the foot is infected, the symptoms show up quickly as scaling between the toes, along the heel borders and along the longitudinal arches with itching, maceration and wrinkling or peeling skin. The definite sign of athlete's foot is tiny blisters that appear in groups and on breaking open leave circular, shiny, red areas underneath. The remedy for this is to use an antifungal medicine with DMSO as the carrier. Since DMSO is itself a fungistatic agent, it discourages the growth of parasites besides allaying the other symptoms.

- **Foot odour.** Scientifically known as **bromidrosis**, this extremely annoying problem is neither infectious nor painful and is not necessarily the result of being unclean. The main cause is a functional disturbance in the nervous system. Besides the obvious odour, symptoms are sogginess between the toes and tenderness of the flesh of the foot. Small blisters may also sometimes appear on the balls of the feet or on the heels. DMSO does check the smell but the applications have to be kept on for long periods.

- **Dancer's foot**. This is an inflammation and, in severe cases, a displacement or fracture of the two small bones located beneath the head of the first metatarsal. These **sesamoid** bones are located in the tendons that run along the joint of the big toe and their function is to lessen the friction when the tendons move. As the name implies, it occurs mostly in people who dance a lot and is characterized by pain and inflammation. DMSO with ultrasound is an excellent form of physical therapy.

- **Metatarsalgia**. Also known as Morton's toe, the symptoms are a sudden sharp and stabbing pain felt in the toes which results from the inflammation of the nerve between the third and fourth metatarsal bones. The inflammation occurs when the head of one displaced metatarsal presses against another and pinches the nerve between them. Every step induces rubbing, pressing and irritation which consequently enlarges the nerve with a coating of scar tissue which forms as a protective cover. The tissue will eventually turn into a **neuroma** that will have to be surgically removed for relief from pain. A DMSO-anesthetic combination injected intramuscularly helps tremendously. Not only is pain relieved, but relaxation of the metatarsal bones also occurs which does away with the need for surgery.

- **Fallen arches**. In this case, because of weakened feet and over-straining, nearly all the bones of the foot change position. The heel bone rolls inwards, the ankle drops, the shin becomes more prominent, the cuboid bone is forced outward and both the big and fifth toes rise. As other bones sink, the inner part of the longitudinal arch “falls”. Classic symptoms are pain and burning in the foot and tiredness with severe ache in the legs. Because the front metatarsal also falls, there is every likelihood of painful calluses growing on the balls of the feet. DMSO has proved highly efficacious for acute foot problems.

- **Heel spurs**. On the heel bone, or calcaneus, at the point where the muscles and ligaments of the foot are attached, one or more spur-like growths of calcium may sometimes form. These are generally padded by the flesh, so it may be years before they turn into a problem. This overgrowth of calcification is brought on by weak feet, prolonged standing, improper footwear or some structural misalignment in the foot. The pain felt in the heel spur, when weight is placed on the bone, does not actually emanate from the spur point but from the inflammation around the muscles where these are attached to the bone. DMSO extracts the pain from the inflamed areas.

- **Ankle sprains**. This common injury occurs when the foot accidentally suffers a violent twist which tears the ligaments that connect the ankle and shin bones. When the sprain is slight, the ankle may just become tender and sensitive. However, with a severe strain the ankle normally becomes swollen, tender and hot with a red or blue discoloration. As the swelling increases it is accompanied by a throbbing pain. DMSO will not only reduce the swelling but will also take away the pain from the inflammation.

DIABETES. DMSO used topically lessens the need for insulin in one out of four persons with juvenile diabetes. It also plays an important role in lessening the incidence of diabetic neuropathy which is a classical sensory nerve function loss frequently seen in elderly diabetics. DMSO given intravenously improves the peripheral vascular blood supply of diabetics by dilating smaller vessels of the lower limbs.

NASAL SINUSITIS, TIC DOULOUREAUX AND HEADACHES. Placed directly into the nostrils, DMSO opens blocked sinuses in a few minutes. It manages this by its ability to cross all membranes of the body without destroying tissue integrity. For tic douloureaux (involuntary repeated contraction of the trigeminal muscles in the face causing excruciating pain), topically applied DMSO brings relief within four days but it is faster and more lasting when injected directly into the spasmed muscle's trigger points. Employed topically to the forehead and the back of the neck, DMSO will bring relief from headaches within thirty minutes. For those who suffer from **temporal arteritis** (superficial pain in the arteries of the temple) DMSO should be used along with cortisone.

SKIN DISEASES, ULCERATIONS AND HERPES. Skin problems such as **pityriasis rosa, tinea versicolor** and **eczematoid dermatitis** clear up with topical applications of DMSO. Additionally, there will be rapid healing of chronic varicose ulcers, diabetic ulcers and ulcerations related to **mycotic infections**. DMSO painted onto the painful lesions connected with **herpes zoster** (shingles), and all other types of herpes, will remove their neurological symptoms at once. It also works well when put on the cold sores of herpes simplex as well. There is no specific therapy available for the treatment of diabetic perforating ulcers. In diabetic patients having **peripheral neuropathy** (a loss of nerve sense at the extremities), perforating foot ulcers are a major problem. DMSO applications completely heal these ulcers within four to fifteen weeks of daily treatment. In the treatment of skin ulcerations, festering wounds and other skin lesions that have resulted from second or third degree burns, the double-quick healing power of DMSO is truly astounding.

CATARACTS AND OTHER EYE PROBLEMS. An ophthalmologist has developed a formula for dissolving cataracts without surgery by combining DMSO with **superoxide dismutase** (SOD). The formulation works quite effectively against glaucoma as well. The discovery that DMSO was also effective in the treatment of retinal disease was made quite accidentally. The retina is a delicate film that covers about two-thirds of the inner surface of the eyeball and is closely attached to an underlying area known as the **choroid**. It is this part of the eye that is sensitive to light. Some patients, who were already afflicted with a retinal disease called **retinitis pigmentosa**, were undergoing treatment with DMSO for certain musculoskeletal disorders. During the course of this treatment, all of them sensed that their vision had somehow improved and reported this fact to the doctors. After one of these patients had a spectacular recovery of vision, ***investigations revealed that the 'culprit' responsible was actually*** *DMSO!* From then on, dimethyl sulfoxide has become the staple and most effective medicine in the treatment of every possible eye disorder or disease.

THE ASTHMATIC SYNDROME. When DMSO is used as a solvent for an anti-inflammatory steroid, an antihistamine preparation of recognised efficacy and a strong **bronchodilator**, and the combination is injected intramuscularly, good to excellent results are achieved in the treatment of asthma.

PREGNANCY AFTER TUBAL OBSTRUCTION. Like other drugs to be avoided during pregnancy, DMSO should probably not be used. However, for women desiring to conceive but who are sterile because of inflammatory tubal obstructions, considerable success has been achieved with the **DMSO hydrotubation** treatment. This particular treatment involves making a solution of DMSO, **chloramphenicol, dexamethasone** and **chlorpheniramine**. This solution is then injected by means of ***ascendant hydrotubation*** through a ***canula.***

THE TREATMENT OF LYME DISEASE. Lyme disease is an infectious disorder that causes joint pains and inflammation and possible heart and nerve complications. The disease was first recognised in 1975 when a cluster of patients in the American town of Lyme, Connecticut, developed a strange type of arthritis. It is contracted and spread by bites from the deer tick, an insect about the size of a dot. Symptoms of Lyme disease include: rash at the site of the tick bite that forms a raised, red circle with a clear centre; flu-like symptoms such as headache, fatigue, fever, chills, sore throat and muscle aches; arthritic complications including paralysis (on the face), abnormal skin sensation and sensitivities, insomnia and hearing loss; heart complications with irregular, rapid or slowed heartbeat; chest pain, fainting, dizziness, and shortness of breath, plus the psychiatric difficulties of depression and dementia.

The topical, oral and intravenous usage of DMSO for most of the troubles entailed in Lyme disease, offers comfort if not permanent relief. Lyme disease is effectively treated with DMSO which *kills* the **spirochete** form of the bacterium, **Borrelia burgdorferi**, which is spread by the bites of this tick. Using DMSO infusions permits the dosage of antibiotics to be cut by more than half when treating Lyme disease because it increases antibiotic effectiveness three-fold. Lessening the dosages also results in avoiding the side effects normally brought on by antibiotics. DMSO carries the antibiotics across the blood-brain barrier as well which prevents psychological complications that include depression and dementia.

MISCELLANEOUS USES OF DMSO. Every system in the body, most physical disorders and many mental dysfunctions are beneficially affected through the use of DMSO with the results varying between "greatly improved" and "completely cured". Some additional uses have come to light recently such as:

- DMSO has cleared up gum conditions and is very effective in cases where decay has reached the dental nerve. In the latter instance, it attacks and clears up both the infection and the inflammation while also relieving the pain.
- DMSO has speeded up the healing process and removed pain in the treatment of "dry socket". The dry socket is an unhealed wound at the site of a tooth extraction. The wound gives intense pain accompanied by pus discharge and sequestra.
- Women with chronic cystic mastitis, a painful breast condition, obtained great relief within a month by topically applying DMSO .
- Conditions of the urinary tract that have **NOT** responded to any other medication have positively improved with DMSO which has corrected urethral blockages as well as reduced pain.
- DMSO, which speeds up blood flow by causing vessels to dilate, has proved to be effective in heart attacks or angina pectoris by preventing damage to the heart muscle.
- Gangrene responds positively to intravenous DMSO. Patients with gangrenous limbs and exposed bone are treated with topical applications of DMSO cream into the local area of trauma. This has proved exceedingly useful in eliminating pain and promoting

tissue healing. Daily infusions of DMSO, carrying trace minerals and zinc plus ascorbic acid, promote a full recovery of the gangrenous state and such healing usually takes place within two weeks.

- The most effective treatment for lead and mercury poisoning is chelation therapy combining DMSO with EDTA. These two toxic metals are capable of entering the brain to lodge in its cells. DMSO and EDTA in the same infusion pull these metallic poisons out of the brain cells.
- One other condition that responds amazingly well to DMSO, with a high vitamin plus minerals solution, given intravenously is **candidiasis**. Patients suffering from the yeast syndrome can have their symptoms wiped out within six weeks.
- Ringing in the ears (tinnitus) and fungal problems disappear unbelievably fast. DMSO can get rid of ear ringing within three IV treatments.

It Can't Be Only Benefits All of the Time?

Unfortunately for DMSO's detractors, that's just how it is—only benefits ALL of the time! It has been proved beyond any doubt that DMSO is totally non-toxic—even in some cases where patients were overdosed! As of publication date, absolutely **NO** serious deleterious reactions have either been noticed or reported. There certainly are a few minor side effects, which are far outweighed by DMSO's benefits, but no one can substantiate any claims to either toxicity or ill health arising from its use. DMSO seems to affect each individual differently. The severity of side effects varies so much across the board that it is extremely difficult to pinpoint any as the side effect. Many patients have firmly averred that even with mild side effects present, these are worth the overall improvements that DMSO brings. The most commonly noted, and reported, side effects are mild indigestion, nausea, stomach cramps, heartburn, soreness and pain at the infusion site, occasional redness and skin irritation, burning sensations, outbreak of rash or hives and an unpleasant odour. All these symptoms either are reduced drastically or vanish completely, without any debilitating physical effects, if the rate of the infusion is reduced, or the infusion is stopped for a while, the solution concentrate is lessened or the amount of DMSO is reduced. These effects are entirely temporary and do not inadvertently resurface at a later date. Some patients evince absolutely none of the side effects while others may show a few symptoms and—of course—there are those who 'insist' on going through the entire gamut!

By far the most unpleasant of these side effects is the odour resembling garlic/oyster that occurs when DMSO is used. However, this harms only human sensibilities—no harm at all is done to the human physique! (And strangely enough, the odour is not noticeable to the person taking the DMSO!) There is no effective method of reducing this odour although in some cases it has been masked by using scents and cosmetics. In cases where DMSO is orally ingested, the odour can be modified if the patient consumes cheese, ice cream or other dairy products before the treatment. Raw milk seems especially able to reduce the mouth odour produced when DMSO is being drunk. Unlike some other drugs, DMSO's stay within the body is extremely limited. Since it is such a good solvent and is capable of rapidly crossing tissues, membranes and blood without harming anything, it does its 'job' and is excreted from the body just as fast. DMSO has definitely opened up a way to control practically all infections

Something Good - At Last!

There are hundreds of patients with various ailments who have been successfully treated with intravenous as well as topical applications of DMSO in both India and abroad. In addition, the *DMSO-hematoxylon* formulation, along with certain herbs, has been successfully used in treating cancer patients. At the time of publishing, all the patients have shown a positive regression of the disease and are progressing satisfactorily without any health complications.

CHAPTER 7

LIPID REPLACEMENT THERAPY - NATURE'S AGE 'REVERSER'

A Chest-Full of - Trouble?

In October 1997, 51-year-old AC came to us as a 'last-ditch' option in his, so far, unsuccessful quest for a normal lifestyle and ***any*** kind of cure. Diagnosis had confirmed triple vessel disease with severe angina pains as 'the icing on the cake', and the biggest 'bonus' of all was the fact that he was declared ***unfit for bypass surgery*** and would have to live with TVD for the rest of his life! He couldn't walk and the lightest exertion caused the angina to flare up agonizingly. Besides other essential heavy drugs prescribed, his desperate clutch at the lifeline also involved swallowing at least 8 to 10 sorbitrate tablets a day. Blood biochemical tests revealed a definite cause for concern. Not only were his lipid profile readings totally unnatural, but they were also alarmingly out of whack! He was immediately put on EDTA chelation and his vital signs were closely monitored during each treatment. Five sessions later the patient complained that he still felt very unwell and that the chelation didn't seem to be working. A further blood test indicated no improvement in the lipid profiles but treatment was continued. After ten sittings, with the patient continuing to complain and the lipid profiles showing no indications of changing for the better, it was decided that the therapy should be discontinued as nothing useful was being achieved. However, on the patient's insistence, treatment was continued for another 20 sessions. Subsequently, with absolutely no positive results forthcoming, therapy was stopped completely.

To say that we were puzzled would be a total understatement! Our other patients, with exactly the same complaints and similar symptoms, were well on their way to recoveries, and we were unable to fathom exactly why EDTA simply '*refused*' to help out poor AC. I went to the USA in early 1998 for a possible answer. I sought out and consulted with many chelation doctors to root out an explanation as to why EDTA hadn't worked on AC. All that I ascertained was that the chelation physicians in the US attained about 70 percent success rates for those patients who strictly followed the prescribed protocol and about 35 to 50 percent successes for those following individual protocols. On the flight from New York to Vienna for a convention, I met a Jewish doctor who made the startling revelation that successful research and tests had been carried out on phospholipids containing **phosphatidycholine** by using rats in experiments pertaining to **lipid replacement.** Interacting with my new friend almost on a daily basis, as we were staying in the same hotel, I eagerly accepted an invitation to Jerusalem to view the work. In Jerusalem, a lot of the research material was made available for perusal as was the added information that the new substance was being manufactured in Switzerland. A trip to Switzerland was made to meet Dr. Sam Baxas, the doctor who had already developed **Plaquex** and had tried out the new drug on twenty human patients—one of whom happened to be his own wife—suffering from irretractible angina pectoris pain (***exactly the problem that plagued AC***!) and who had so far been sustained on beta blockers and nitroglycerin products. The Swiss researchers were absolutely elated when, after completing the full course of treatment, 19 of the 20 patients were declared asymptomatic ***and totally free from reliance on any medication whatsoever***. (One patient had dropped out of the programme during the initial stage of the treatment.)

Immediately upon my return to India, with both Plaquex and the new protocol firmly in hand, AC was contacted and started off on the new therapy. It should be emphasized here that at the time of re-commencing treatment, the patient's lipid profiles still continued to be totally haywire—especially **lipoprotein(a),** with readings of **140** as against a reference range of zero to 30.0, and the ratio of **total cholesterol** to **HDL** reading **12** as against a maximum permissible value of 5! His individual **cholesterol** and **triglyceride** readings were also extremely volatile making him a very high-risk patient. Due to his delicate physical condition, initially he was given only one plaquex infusion per week. After five sessions, a subtle difference in his overall demeanour became quite evident, with AC himself asserting that he had started to feel better. Since there was a definite physical improvement, treatment sessions were increased to twice a week. After eight plaquex treatments, a biochemical blood profile was carried out which revealed that the *lipoprotein(a)* reading had dropped from 140 to **69** and the *cholesterol/HDL* ratio from 12 to **7.** Following the eighth plaquex treatment it was decided to give him one session of EDTA infusion. In all, AC underwent a total of 30 treatment sessions of which 25 were with plaquex interspersed with five sessions of EDTA. Toward the end of the therapy, *lipoprotein(a)* had further dropped to **39** (we were unable to bring it lower than that) and the *cholesterol/HDL* ratio was completely normalized. Treatment was discontinued after the thirtieth session as we estimated that the maximum possible improvement had been achieved. The patient has since resumed all his normal activities and regularly takes a daily, brisk 45-minute walk. Coupled with some essential lifestyle changes, he has been following a strict diet regimen and is in the process of enjoying his resurrected health. Following the positive indications of our experiments with the **EDTA-Plaquex** combination, we have thereafter used this 'Combination Therapy' quite extensively with an ***amazing 98 percent success rate!***

One More Example

In July 1985, JK, who had six months earlier been diagnosed as suffering from angina and hypertension, was admitted to a hospital when he experienced chest heaviness. Though he was a non-smoker, non-alcoholic and non-diabetic, his BP registered 160/100 with a pulse rate of 80 beats/min. Investigations revealed a calcified lesion in the left upper zone of the heart and the EKG showed inferior wall ischemia plus an infarction. Other tests proved just as bad, so on July 20 bypass surgery was conducted. He was discharged from the hospital two weeks later in stable condition and with the standard medication prescribed as post-operative therapy. He took his medicines regularly and continued having periodic check-ups. For five years things went smoothly but in January 1990, JK had to be hospitalized again when he complained of uneasiness, heaviness and chest pains. When a tread mill test (TMT) showed complications, he was referred to a neurologist who carried out further tests and recommended that he undergo angioplasty. Since the procedure was relatively new at the time, JK declined the treatment and sought out a cardiologist who would treat him only with medicines. The medications appeared to have helped as some of the symptoms ***were*** relieved.

In September 1993, JK was hospitalized once more because of angina. His pulse rate had shot up to 90 and his BP was 170/90. This time the TMT revealed reversible ischemia and other tests confirmed partial arterial blockages. He was again advised to undergo angioplasty but preferred to carry on with medicines as he felt that the surgical procedure would not prove successful. In November 1998, JK developed chest pains while resting, which necessitated hospitalization and this time the diagnosis revealed a very serious condition. He had ischemic heart disease, unstable angina, multiple vessel disease, severe left ventricle disease and a past myocardial infarction. JK had no choice this time but to submit to angioplasty and stenting. Balloon angioplasty was carried out in three places and two intracoronary stents were placed in the arteries. He was stable when discharged from the hospital, but a few days later unstable angina reared its ugly head again. Five months later JK was back in the hospital with his condition bad enough to warrant two further angioplasties and the fitting of one more stent. All these procedures were to no avail because his angina refused to 'go away' and he had reached a stage where he was almost totally bedridden.

JK was 63 years old when he came to us in May 1999, after having been told by the hospital that nothing further could be done. Besides having chronic unstable angina and being unable to walk without severe chest pains, he also had a host of other complaints that left him completely debilitated. From June onwards we commenced plaquex therapy alternating this with sessions of chelation therapy. After the nineteenth treatment session JK developed mild arrhythmias and had to be put on drug medication for two weeks to overcome these symptoms. A fortnight later he was back on plaquex/chelation and showing marked improvements in his overall condition. After the thirtieth session JK underwent a realistic geometric cartiography imaging test (RGCI) which indicated enhanced cardiac functioning, normal blood supply and hardly any evidence of disease. After 20 plaquex and 14 chelation (34 sittings) sessions, JK was able to walk quite normally for about an hour but had to rest for five minutes after every 20 minutes of walking. His cardiologist, with whom we had been closely coordinating treatment, evaluated JK's condition and was very happy with the progress made. After the fortieth treatment, the cardiologist drastically reduced essential medication since the patient had achieved remarkable improvement in his general condition. Therapy continued for a total of 55 sessions. JK is now off almost all medication, is able to take brisk daily walks unaided and without discomfort, and is also able to attend his shop from 11 AM to 9 PM every single day.

Much Ado Being Made About 'Something'

And rightly so! The discovery that **Lipid Replacement Therapy**, or **Plaquex Therapy**, is a positive age-reverser definitely merits the entire quota of hue and cry! It may appear somewhat odd to label 'ageing' as a disease since it's a perfectly normal and natural sequence of life. However, after listing all the degenerative effects of ageing on the human body, one most certainly should refer to the process as a '***disease***'. For a better understanding of what actually happens, we will once again return to that ubiquitous little particle in the human body *–the cell.*

We start our trek through life generally blessed with healthy and young bodies. As we 'age' with time, the trillions of cells inside us undergo various changes which also trigger many alterations in the membrane of each cell. Youthful cells will naturally have membranes with youthful characteristics—***those properties and attributes which perform youthful cell functions.*** Though within us there are a countless number of cells, comprising the tissues and organs, there are actually fewer than three hundred types—each contributing to the our 'collective' life by its special and singular function. The ability of a cell to function greatly depends upon the function of the membranes that encase each cell and its internal structure. Membranes are complex constructions comprised of *lipids*, carbohydrates and proteins as well as receptors, channels and enzymes. There are a variety of these fat-based *lipids* which form a bi-layer and occur in very specific ratios. This bi-layer is the template, or molecular pattern, governing assembly of protein, etc. for all membranes and provides mechanical stability, flexibility, electrical resistance and other characteristics. Membrane activities are brought about through and dependent upon preferred composition ratios of various lipid components. The ratios are age-sensitive and influence membrane-based proteins which include enzyme, transfer, receptor and channel functions. *Ageing* causes the lipid ratios and structures of the membranes to change and degrade, resulting in the cell slowly losing its youthful elasticity, shape and function. Consequently, the membrane performs as a 'stumbling block', neither admitting nourishment into the cell nor permitting waste matter to leave. Over time, these kinds of changes to and activities of the membrane cause the cell to lie in a fallow and unproductive state, eventually causing total failure, some earlier and some later, of all cells, tissues and organs. ***In essence, the capacity and competence of the cells to function youthfully deteriorates with ageing.***

Plaquex Therapy reverses these age-related changes in the lipid ratios of the cell membranes, bringing back the readings to a ratio that is closer to what existed when the individual was younger. As the membranes go through a revival, receptor, enzyme and channel functions resume and the cell changes back to an earlier shape and more youthful function. It's almost as if the cell is reawakening from an extended sleep. With the membranes functioning correctly once more—with the admittance of nourishment and the excretion of waste—the cell gradually regains its lost elasticity. A reversal of age-related changes prevalent in older people is possible through the utilization of this therapy, while for younger people the ***onset*** of age-related changes can definitely be deferred.

What Are the Therapeutic Applications?

It has been positively established that plaquex therapy has the following therapeutic applications:

1. Reverts cellular, tissue, organ and bodily functions to a younger state.
2. Retards or prevents the progression of ageing irrespective of chronological age.
3. Diminishes, reverses or prevents the risk of atherogenesis.
4. Enhances the capacity for cardiorespiratory stress.
5. Restores immunocompetence.
6. Enhances sexual competence.
7. Improves impaired renal function.
8. Reduces high blood pressure (both elevated systolic and diastolic).
9. Reverses myocardial hypertrophy (enlarged heart)
10. Reduces cholesterol and cleans up all arteries of the body.

But How Does All This Happen?

A change in the membrane lipid composition is one of the biochemical changes that take place with ageing in human plasma membranes. The main variation occurs in the relative composition of **phosphatidylcholine (PC)**, **sphingomyelin (SM)** and **cholesterol (C)**. **PC** ***decreases*** with age while both **SM** and **C** ***increase*** with age. The changes in the relative amounts of **PC** and **SM** are particularly great in tissues which have low **phospholipid** turnovers. As ageing progresses, the plasma membranes associated with the aorta and arterial well show a substantial decrease in the **PC/SM** ratio. SM shows an upward surge in several diseases, including atherosclerosis, and its content can be as high as 70 to 80 percent of the total phospholipids in the case of advanced aortic lesion. The two important factors that differentiate PC and SM lipids are (1) the phase transition temperature of the lipids and (2) the hydrogen bonding character of the two in a lipid bi-layer. PC transitions at a much higher physiological temperature range. Where hydrogen bonding is concerned, because of the difference in the polar regions of these two lipids, PC can only serve as a hydrogen donor whereas SM is enabled both as a receptor and a donor. It is not known if these differences are the main cause, but what is known is the fact that the relative content of PC to SM in plasma membranes significantly affects cell functioning. Not only, as already stated, does ageing decrease the PC/SM ratio, but the lipid composition changes are also accompanied by dramatic changes in heart cell activity as well as significant increases in the activities of at least seven enzymes. One of these enzymes is **creatine phosphokinase** (**CPK**) which plays a major role in intracellular energy transport from the **mitochondria** to the **myofibrils** as well as the regulation of energy production coupled to energy utilization.

The ratio of cholesterol to phospholipids is also an important determining factor in regulating the properties of biological membranes. Certain properties of the membranes can be altered to become enriched by depleting the C content. There is a strong positive correlation between changes in SM and C levels in plasma membranes, in that, *changes in the content of one are followed by changes in the other*. Plaquex has the ability to alter the lipid composition of the biological cells and restore the original composition through lipid exchange. This is an extremely important aspect where ***increasing*** both PC/SM and PC/C ratios are concerned— prime factors in ***deferring the ageing process***. Lipid exchange also brings about a reduction in cellular enzymes, such as CPK, which have the normal tendency of increasing with age. PC- rich **liposomes** are able to reverse age-related changes in the lipid compositions of the heart muscle besides dramatically improving the ability to withstand respiratory stress. Another quality of plaquex is its ability to cleanse all the arteries in the body and bring about a dramatic improvement in both cerebral function and the return of sexual potency. Of course, with the return of normal circulation to the coronary vessels, a disappearance of debilitating symptoms is also evident.

When Was the Lipid Replacement Theory Discovered?

From time immemorial doctors, scientists and philosophers have been hard on the heels of a possible cure and reversal of the ageing process—without much success. Ancient Indian texts have mentioned certain herbal and traditional therapies that successfully combated the ageing process. However, information on the methods and medicines used has disappeared into the mists of time. Modern speculation proposes that our ancestors lived longer because they led healthier lives and better understood the laws of nature. In the present, *atherosclerosis* has long been recognised as a feature of ageing. It has garnered the most interest and given rise to extensive medical attention since its effects are not only medically and economically widespread, but also degrading and quite devastating. It was only after World War I that doctors eventually understood how certain dietary restrictions could slow the progress of atherosclerosis, but that nothing could be done about the lasting damage inflicted upon cellular membranes, the cells and their functions. In 1950 research was able to establish that infusions of *phosphatidylcholine* could displace cholesterol from the cells into the bloodstream, ***or reverse cholesterol transport***, thereby eliminating the excess cholesterol through excretion. In experiments on rabbits it was noted that in cases of artificially induced high levels of cholesterol, PC infusions successfully reversed the atherosclerotic process. Further studies showed that humans could safely undergo the therapy and that positive results would also be achieved. A more recent finding of medical research is the effectiveness of PC infusions in the reversal of age-related changes in cells; how they increase the PC/SM ratios and how they restore not only the youthful beating rates of myocytes, but also their young spherical configurations. The latter remarkable occurrence seems to be related to the action of two intracellular protein fibres, **actin** and **myosin**, which are capable of contracting and relaxing, thereby maintaining not only the cellular shapes but also being able to move particles about within the cell or even move the very cell itself.

To confirm exactly how reversal of age-related changes were initiated and undertaken, all avenues pertaining to the nature and extent of damage to the individual cell, as well as to the cellular membranes' properties and functions, were explored in depth. In 1952, studies showed that **PC** infusions could alter both cholesterol content and atherosclerotic progression, but it took three more decades before acceptance was finally accorded to the fact that ***intravenously administered liposomes were effective as synthetic anti-atherogenic***

lipid particles. It was also confirmed that reduction in **C** level alone, whether in plasma or at the cellular level, is not enough to reverse age-related cellular changes. To achieve this aim, alterations are required in the **SM** composition of membrane lipids or a reversion to a younger **PC/SM** ratio. It was only in 1989 that **Lipid Replacement Therapy** officially gained recognition. The first human treatments commenced in 1992 and were an unqualified, resounding success. Those first patients are leading healthy, medically uncomplicated lives even today.

A Short History of Plaquex Clinical Studies

It was some years ago that physicians stumbled onto this generic product isolated from soy beans. This had already been clinically tested about 30 years prior to its utilization as a treatment for patients severely ill with coronary artery disease. As mentioned earlier in the chapter, 20 patients suffering from irretractible angina pectoris pain, who were being sustained on beta blockers and nitroglycerin products, were entered into a treatment programme which entailed the infusion of a solution made from this generic product over a period of 50-60 minutes, three to five times per week for a total of 30 treatments. To the amazement of the research team, all 19 of the patients who completed the programme became totally asymptomatic and free from reliance on any medication as soon as the treatment protocol was completed. Based on these results, another medical research team in Switzerland initiated their own clinical study on a small group of severely ill patients. Two of these had already undergone two bypass operations each and had been sent home to die; one patient had two angioplasties behind her but continued to suffer from angina pectoris and a fourth patient was documented by a thallium stress test to have heart perfusion problems. The results on the bypass patients after 30 treatments with this protocol were unbelievable. Both of them became more mentally alert, sexually active, completely asymptomatic and free of the need for any medication. Follow-up thallium stress tests showed normal perfusion studies of the heart. The female patient with two angioplasties showed the same results—completely asymptomatic after the series of treatments and totally off all medication. A follow-up thallium stress test at the University Hospital in Basel, Switzerland, showed her to have a completely normal perfusion of the heart. The fourth patient also underwent a remarkable cure.

Although the medication was developed over 50 years ago, its exclusive use for the purpose of breaking down arteriosclerotic plaque, in all of the arteries of the body, is relatively new. Researchers continue to be amazed at the efficacy with which plaquex tackles severe angina and coronary blockages. Those patients who complete the treatment protocol normally revert to an asymptomatic state and, in many cases, may no longer need the life-sustaining heart medications that were originally prescribed! And best of all, ***plaquex therapy produces absolutely* no *side effects or ill-effects.***

Because of its ability to cleanse **ALL** the arteries in the body, patients on plaquex will notice a dramatic improvement in cerebral function, a return of sexual potency and, of course, the return of normal circulation to the coronary vessels and other vital organs followed by the disappearance of adverse symptoms. ***The only indications are positive ones*** and patients become aware of the vast improvements even while in the process of undergoing treatment. The long-lasting effects of plaquex are quite evident from the fact that the first patients to receive this treatment in 1992 are doing well and show no signs of needing further therapy. There are many patients all over the world who have been, and continue to be, successfully treated with this therapy. Not only have a majority achieved total cures but also, amazingly, ***no side effects have been noted to date in even a single patient!!***

What Is the Treatment Protocol?

Prior to treatment, patients are required to undergo some essential tests such as stress, cardiac catheterization and, with the MRI equipment now capable of imaging the coronary vessels, to check for partial or incomplete blockages. Tests should be repeated after completion of the therapy as well. Levels of ***homocysteine***, ***B6***, ***fibrinogen***, ***cholesterol***, ***triglycerides*** as well as ***HDL*** and ***LDL*** should also be tested both before and after treatment. Patients may continue with their medications such as beta blockers, nitroglycerin and anti-hypertensives. However, ***cholesterol lowering drugs should be* stopped *during the course of the therapy***. Patients may continue with their ***coumadin*** treatment, if any. As treatment progresses, all medications can be gradually reduced and, as has generally been noticed, by the end of 20 treatment sessions the only requirement may be nitroglycerin on an "as needed basis". More severe cases could require between 30 to 40 therapy sittings before they eventually recover to an asymptomatic state with no further reliance on medication.

Sometimes there may be indications that a combination of plaquex and EDTA would be more effective for certain stubborn situations, however, it must be remembered that ***plaquex cannot be mixed with EDTA solution***. Through experimentation we have found that the combination, if required, works best when plaquex therapy is *alternated* with chelation therapy. The most effective results are obtained when a session of chelation administration is done after two or three plaquex sittings. Even having them in alternating sessions is efficacious. What cannot be denied is that the combined effects of chelation (with its properties of removing heavy metals plus its anti-oxidative qualities) and plaquex (with its vessel cleaning qualities) provide an ideal treatment modality. In the normal administration of *lipid replacement therapy,* the patient is given plaquex in 5% dextrose intravenously over a period of 60 to 70 minutes, two to three times a week. The number of treatments could vary between 20 to 30 sessions depending upon the medical analysis and after evaluation of test findings and reports. All treatments are carried out on an out-patient basis.

Summing Up the Good Effects

Heart and Circulatory Diseases. The recognised factors for these age-related diseases include the accumulation of excess cholesterol in the human cell membrane, a condition which has never before been medically modified, and the presence of a substance known as Lipoprotein(a) – or just simply **Lp(a)**. This is a substance that behaves as the worst type of cholesterol. The damaging effects of Lp(a) have only now been widely recognised in international medical circles. Plaquex not only succeeds in removing excess cholesterol from the membranes of the cell, ***a feat never before medically accomplished***, but it also substantially and effectively lowers Lp(a) and in so doing ***does what no other therapy in the world has ever done before***. Lp(a) is a critical factor in the formation of blood clots that are the cause of myocardial infarction and strokes. It is associated with the narrowing or **restenosis** of (1) the transplanted vein in coronary artery bypass surgery and (2) the artery that is expanded through angioplasty. Restenosis occurs in more than 30% of such invasive procedures, usually within several months of the event.

Lipid Replacement/removal. Plaquex accomplishes (1) removal of excess lipids, such as **sphingomylin**, that have accumulated in the cells and cell membranes as the person ages and (2) restoration of

phosphatidycholine, another essential membrane lipid, which gets depleted with ageing.

Reverse Cholesterol Transport. Cholesterol is transported in the bloodstream, and from cell to cell, by certain carriers called **lipoproteins**. Plaquex, once administered into the body, creates a new lipoprotein which functions as a surrogate high density lipoprotein (**HDL**) or the *good* cholesterol. It aids in the removal of excess cholesterol from the cells' membranes and, together with the body's other HDLs, transports the large pool of free cholesterol—from their deposits in blood vessels, fat cells, muscle, and connective tissue—to the liver from where it is finally excreted. The benefits of reverse cholesterol transport and the significant lowering of Lp(a) have the capacity to inhibit the progress of atherosclerosis and/or promote its regression. Plaquex's ability to clean all the arteries of the body is instrumental in granting its recipient a high level "Quality Of Life".

What Is This "High Quality of Life"?

In December 1995, SRB suffered a myocardial infarction and the angiography revealed that his was a severe case of coronary artery disease with multiple plaque blockages and a dysfunctional blood flow. He was prescribed heavy medication to overcome the major symptoms and was advised to undergo bypass surgery if the medicines proved ineffective. Though the doctors pressed for immediate surgery, SRB refused and was quite content to continue treatment purely with medicine. The medication appeared to have some positive effect as SRB did not have any overt complaints. Almost six years later he was once again admitted to a hospital with complaints of left-sided chest pain and on the same day experienced a fainting attack with transient loss of consciousness. For the next two to three days he also suffered loose motions, fever and vomiting. Another angiography showed that the conditions of artery blockages, dysfunctional blood flow and calcification had considerably worsened. He was once again urged to have an immediate bypass, the cardiologist cautioning him that if he didn't have the operation he was likely to die within the next three days! SRB stubbornly refused to give in, preferred to stay on medicine and had himself discharged.

Two weeks before his second admission to the hospital, SRB had approached us for an alternative means of curing his problem. He was suffering from coronary artery disease and angina, triple vessel disease, hypertension for the last ten years, diabetes mellitus that was controlled by diet and oral medication, bronchial asthma contracted before 1990, breathlessness on walking for more than five minutes or if climbing stairs and occasional pain in the left shoulder that came and went by itself. In other words, SRB was a total physical wreck and the age of 56 had nothing to do with it! As soon as he got out of the hospital we commenced plaquex therapy on SRB. At the time, his serum creatinine level was 1.8 mg/dl as against the permissible range of 0.6 to 1.4. After eleven therapy sessions, clinical blood tests showed that sugar, cholesterol and triglyceride levels had all reverted to normal. On completion of 22 sittings, serum creatinine was down to 1.43 and an ECG showed *marked symptomatic improvement*. By the thirtieth treatment, creatinine levels stabilized at 1.1 which is completely normal.

Today SRB leads a totally normal life with no overt debilitating symptoms. His blood sugar is within limits and urine samples also have nil sugar indications. Though he still has to maintain certain essential medication, SRB has resumed all routine activities including regular walks. He has been maintaining a controlled diet, and has changed his life style, i.e., avoiding the 'hurry and scurry' syndrome!

Priceless Good Health

SRB's story is a striking example of regaining "high quality life" after orthodox medicine had literally written him off as a hopeless case! The regenerative qualities of plaquex prop up your body cells, cause them to function youthfully, and also ensure that this enhancement endures. Nothing in this world comes for free, and this is very true where plaquex is concerned. After 'rejuvenation' it is incumbent upon us to maintain the goodness that we have received. If this means giving up certain "good things" of life to preserve health, so be it. The ultimate rewards are immeasurable!

CHAPTER 8

ULTRAVIOLET BLOOD IRRADIATION THERAPY: NATURE'S WASH

Wonders Never Cease!

Strange but true—and definitely so in the case of this therapy. Who would have thought that a cure for many maladies is possible through ***the treatment of oneself by one's own irradiated blood?*** This therapy is known by many names, such as ***Ultraviolet Blood Irradiation (UVBI), Photoluminescent Therapy, Photochemotherapy, Photobiological Therapy, Hematologic Oxidative Therapy, Hemo-irradiation, Photopheresis, Photodynamic Therapy*** etc., and was first introduced in the 1930's mainly to effectively combat the polio virus. **Ultraviolet light** has been used as a disinfectant for many years and most contaminated objects can be cleansed rapidly of viruses and bacteria in this manner. When a small quantity of blood is treated with ultraviolet light, an astounding thing happens. Through a mechanism that probably involves increased production of **lymphokines**, the immune system is activated to "attack" either cancer cells or invading organisms. The exact mechanisms through which this is achieved are not fully known at present.

Light and health have always gone together. However, it is only recently that scientists and doctors are beginning to realize and understand just how beneficial light is to life. The fact that light ***has*** inherent qualities that promote health has spurred the usage of coupling its qualities with modern medical science, resulting in greater beneficial implications for the long-term maintenance of health in general. By making use of the ultra-violet portion of the light spectrum, doctors have found it possible to cure a wide variety of, what were previously thought of as, "hopeless" diseases. The best part of this therapy is the fact that no sophisticated or complicated equipment or drugs are required to put the theory of **photoluminescence** into practice. The key to light therapy's versatility is its ability to cure through the stimulation of the body's own immune responses.

You may be slightly nonplussed as to what exactly "light therapy" is or embodies. As already mentioned, it is known and acknowledged under many headings. These names are instructive in themselves, as they definitely give us an idea of the diversity of the research that has been carried out. However, the most appropriate name would be **photoluminescence** as '*photo*' refers to light and '*luminescence*' means emission of light. Although we are mainly concerned here with ultraviolet light, one should not overlook the fact that the energies from x-rays through infrareds can also be applied to the treatment of disease. However, the most dramatic effect of ultraviolet irradiation ***is the stimulation of the immune system and various enzyme systems***. Originally, the method of exposing blood to light was developed to utilize the bacteria-killing property of ultraviolet rays by directly irradiating the bloodstream and thereby successfully fighting bloodstream infection. However, while treating patients for this specific purpose, it was noticed that many additional, and decidedly beneficial, biological and biochemical reactions were also taking place. In 1942, Dr. Virgil K. Hancock and others observed that the following reactions had occurred:

- Inactivation of toxins and viruses.
- Destruction and inhibition of bacterial growth.
- Increase in the oxygen combining power of the blood and the oxygen transportation to the organs.
- Activation of steroid hormones.
- Activation of white blood cells.
- Immunostimulation of cellular and humoral immunity.
- Stimulation of fibrinolysis.
- Decreased viscosity of blood.
- Improved microcirculation.
- Stimulation of corticosteroid production.
- Decreased platelet aggregation.

In 1943, Dr. George P. Miley published his particular work regarding the effects of this therapy on viral pneumonia and the observations were:

- Complete disappearance of toxic symptoms 24 – 76 hours after a single treatment.
- Disappearance of cough in 3 – 7 days.
- X-ray evidence of lung clearing in 24 – 96 hours.

There is no other therapy existing today that can do that. There are many articles that have been written concerning the effectiveness of this technique for treating infectious disease. The added fact that UVBI is approved and being used for beating a particular type of cancer means that we are dealing here with an excellent immune stimulant.

How Did It Develop?

The history of UVBI therapy goes back to the 1880's when Dr. Niels Ryberg Finsen discovered the therapeutic applications of ultraviolet radiation. He didn't actually use it in this manner; he only discovered the potential through his scientific approach to the subject. His work in the late 19th century, especially in the area of ultraviolet treatment of skin and mucous membranes, was responsible for both the medical world getting to know about the healing action of this form of radiant energy and the voluminous research and studies that ensued as a result. By the 1930s, ultraviolet light was being employed as a systematic treatment through external radiation, and its application to specific disease problems began in the 1940s. The move from external irradiation to blood irradiation was a quantum leap in the realization of light's healing potential. When bacteria or fungus or viruses entered the bloodstream or lodged under even the flimsiest layers of tissue, they were generally seen to be impervious to the bactericidal effect of external radiation. Researchers discovered that a small amount of blood irradiated by ultraviolet light could effectively destroy and inactivate these infectious invaders no matter where they were hidden in the body.

One of the pioneers in this form of therapy, and the man who developed the model from which the current equipment is copied, is Emmet S. Knott. In 1928, he irradiated blood from the first human subject—a case of sepsis (blood borne bacterial infection). This patient recovered rapidly. Working together, Knott and Dr. Virgil Hancock presented many of the successfully treated cases to a local medical society in the state of Washington, USA. The first article published on their work was in 1934. By June 1942, 6,520 patients had been treated with ultraviolet therapy. ***It worked in over 95% of the cases without any harmful side effects***—a difficult concept to accept in the light of all the harmful side and after effects of so many of the then current orthodox therapies. All of this marvelous work was done before there was any understanding of the immune system. ***Thus, they made a monumental breakthrough in therapeutics without any understanding of the mechanism.*** The process of performing this therapy is almost the same today as it was then. Essentially, though the exact mechanism is ***still*** to be unraveled, it had been proved beyond any doubt that the proper administration of photoluminescence was an effective method of treatment for cases of life-threatening infection and brought about total recoveries—especially where existing therapies were inadequate. In addition, ultraviolet irradiation completely neutralized even the most virulent of toxins.

Unfortunately, today's orthodox medical following is almost totally ignorant about ultraviolet therapy—few physicians are well grounded in physics, biophysics and technical therapeutic procedures involving ultraviolet radiation, which are essential tools for understanding the basics of the process. Therefore, this wonderfully effective method of treatment currently plays second fiddle when it could possibly be the main assault weapon in cases of life-threatening infection. The neglect of this procedure at present will probably go down as one of the century's great tragedies. A massive number of deaths and untold suffering could have been averted if ultraviolet therapy had not been largely abandoned after the discovery of penicillin.

Photoluminescence is a viable and effective treatment option for many diverse medical conditions. The therapy is recommended for the following conditions:

- Autoimmune diseases: rheumatoid arthritis, etc.
- Cancer
- Emphysema
- Immune deficiency problems
- Inflammatory processes: fibrositis, bursitis, iritis, pancreatitis.
- Non-healing wounds
- Osteomyelitis (polio)
- Peripheral vascular disease
- Pneumonia
- Septicemia
- Thrombophlebitis
- Viral infections (hepatitis, respiratory, etc.)
- Wound infections

How Is It Applied?

The procedure is disarmingly simple , safe, and effective against a broad spectrum of viruses and bacteria. Only a maximum of 50 ml of the patient's blood is treated over a 10-minute period using the accepted therapeutic UV band of light waves. The effect on the blood which is treated spreads throughout the body and clinical improvement is ordinarily rapid. The device itself is safe, non-toxic, easy to apply, and uses no drugs.

Ultraviolet blood irradiation produces a rapid detoxifying effect with waning of toxic symptoms. The concentration of venous oxygen is increased in those with depressed blood oxygen values. There is a rapid increase in resistance to both acute and chronic viral and bacterial infection and no harmful effects have been observed. Diseases that have been successfully treated include viral infections, hepatitis, bacterial infections, hypoxemia, and many others in the category of blood-borne infections. ***Three to five treatments usually inactivate most viruses with no risk, hazard, or pain to the patient.***

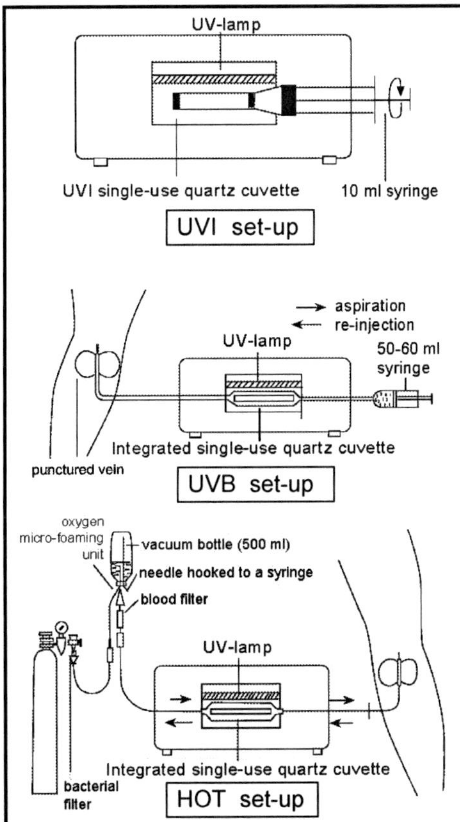

Figure 8 – Blood Irradiation Schemes

There are three procedures for carrying out the treatment. However, the two most common are methods 1 or 2, which are widely employed by physicians all over Europe, the USA and are currently performed at our clinic as well. The three procedures, shown by Figure 8, are:

1. **UVB**: Ultraviolet irradiation of Blood (IV) is done by aspirating and re-injecting 50 ml of venous blood by a 50 ml syringe, pre-filled with 10 ml of sodium citrate, through a quartz cuvette that

is integrated in the tubing and placed in the UV generator as shown by the adjacent figure. The entire procedure takes about ten minutes.

2. **HOT**: Hematogenic Oxydation Therapy (IV) is performed after setting up the system with a vacuum bottle, special tubing with glass cuvette etc. (as shown by the figure). Then 20 ml of sodium citrate is first drawn into the bottle followed by an aspiration of 50 to 60 ml of venous blood. The blood is thoroughly oxygenated within the bottle for 2 to 3 minutes till it turns frothy and is then allowed to be re-infused, through the glass cuvette placed in the UV generator, into the vein by gravity. The whole procedure takes only 20 minutes.

3. **UVI**: Ultraviolet irradiation of blood for Intramuscular re-injection. In this method 5ml of blood is taken out in a quartz cuvette syringe and re-injected intramuscularly after irradiation in the UV generator where it revolves for 5 minutes.

Frequency of Treatment. Depending on the condition of the patient and the extent of dysfunction or debility, patients are given treatments twice weekly for three weeks and once a week thereafter. The method of administration can be any of the above three, or a combination. This will be decided upon by the attending physician. A series of six to ten sessions is generally required for the following disorders:

- Organic, peripheral, arterial, venous, cerebral and coronary circulatory disorders.
- Functional circulatory disorders (migraine, Raynaud's disease, etc.).
- Diabetes mellitus.
- Disorders of lipid metabolism.
- Supportive cancer therapy.

What Are the Beneficial Effects?

The many effects of ultraviolet light are elucidated below for your better understanding.

Effect of Ultraviolet Light On the Blood. The number of red cells, white cells and platelets in the blood may increase following irradiation which also produces a lowered blood sugar, an increased sugar tolerance and elevated blood calcium. Although blood cells can be made to increase through ultraviolet irradiation, a lowering of the white cell count can also be induced by using certain light frequencies. It cannot be said with any degree of certainty that ultraviolet radiation definitely increases resistance to specific or general infection. However, what ***IS*** certain is that there exists a definite relationship between sunlight and the general character of disease, growth and nutrition. In simple terms, this means that there is no evidence that by taking the ultraviolet treatment, one can prevent the onset of disease. On the flip side, it has been ***definitely*** demonstrated that treatment with ultraviolet light, ***after*** contracting certain diseases, dramatically decreases both the duration and the malignant character of the illness.

Enhancing Oxygen Exchange Efficiency. A very evident physical reaction to photoluminescent therapy is the rapidity with which the pinkness of the skin increases. This happens because of vasodilation (expansion of the blood vessels next to the skin) and an increased ability of the blood to pick up oxygen. This extra oxygen is picked up from the plasma itself—the non-cellular portion of the blood. There is no doubt that there is an increase in the efficiency of oxygen exchange and oxygen metabolism following UVBI treatment.

Cardiovascular Effects. A lot of research has yet to be conducted on the long-term effects of photoluminescence on the cardiovascular system. It is not known, for example, what effect light therapy has on congestive heart failure. However, the overwhelmingly positive effects of the therapy clearly indicate that there probably ***are*** long-term benefits for such maladies as hypertension and heart failure.

Effects On the Autonomic Nervous System. The vasodilation mentioned earlier parallels many other effects on the autonomic nervous system. The heightened coloration effect takes place a few minutes after UVBI is administered and may last for more than 30 days accompanied by a feeling of general well-being. This is always regarded as a favorable sign no matter which disease is under treatment. Another remarkable effect on the autonomic nervous system is the speed with which paralysis of the bowel can be relieved with UVBI. The smooth muscle tone and contracting ability of the gastrointestinal musculature are restored to normal within 12 to 24 hours after a single treatment session indicating that a portion of the autonomic nervous system controlling the gastrointestinal tract is also restored to normal function. The favorable response of infection and inflammation of the superficial veins is so dramatic that it is hard to believe that the results attained only entail the destruction of the bacterial infection. Allergic asthma also responds positively. All these positive indications point to the fact that the hitherto out of balance autonomic nervous system is brought back to normal functioning through UVBI.

Metabolic Effects. Ultraviolet irradiation makes the body rid itself of uric acid faster, a positive indication for the treatment of gout—which is an arthritic condition caused by the deposition of uric acid in the joints. Irradiation also doubles the fat content of the blood causing cholesterol to increase by as much as 30 percent. This may be seen as a positive action because cholesterol is vital to the body's immune defences. It has been discovered that blood sugar is temporarily decreased by UVBI because the excretion of insulin increases during the therapy. Diabetic retinopathy is effectively treated with UVBI as erythrocyte aggregation is suppressed. UVBI can prevent and cure rickets but is ineffective in the healing of fractures. UVBI also relieves tetany conditions which is a low calcium state that causes spasm of the muscles.

Effects of Physiologic Conditions On Sensitivity to UV Light. One of the indications of sensitivity to UV light is the tendency to get sunburn even with minimal exposure, with the intensity of the effects being different for different individuals. However those with unstable nervous systems, overactive thyroid glands, elevated blood pressure or active tuberculosis have shown significantly increased sensitivity to UV light. Many drugs and antibiotics are also responsible for increasing UV light sensitivity as do some other factors such as the first day of the menstrual cycle, pregnancy and an acid diet. However, alkaline salves tend to increase skin sensitivity to UV light more than acid salves.

Does UVBI Have Detoxifying Effects?

The therapy ***does*** have the most remarkable and dramatic detoxifying effects. One of the many useful aspects of light is its deadly effect on **toxins**. Toxins are poisonous substances which could take the form of animal poison (snake or scorpion venom, rabies etc.) or bacterial poison (strep, botulin, tetanus, plague, gas gangrene, typhoid germ, diphtheria germ etc.). These toxins are quickly deactivated on exposure to light and ***the potential health benefits of such a phenomenon are incalculable.*** This anti-toxin property of UV light has been known for over 100 years.

So completely does light deactivate toxins, that no one actually need die of snakebite or bacterial poisoning if UVBI treatment is close at hand.

Though early research with snake venoms concentrated on the effects of sunlight, rather than UVBI, on toxins, such research definitely had applications to later work with photoluminescence. **Crotalus** venom (found in rattlesnakes, water moccasins, copperheads etc.) contains heavy doses of a toxin called **hemorrhagin** which destroys red blood cells. This venom also contains a small quantity of neurotoxin (a poison that attacks nerves) but this is not a primary effect of this type of venomous snake. When crotalus venom was exposed to sunlight in the presence of a certain photo-active dye called **eosin**, it was seen that the poison's toxicity was reduced almost 200 times! Without the sunlight, though, the dye had no effect on the poison and the toxin could destroy red blood cells in the dark. The intravascular clotting factors (elements in the blood that cause it to coagulate) in crotalus venom were also completely neutralized. However, there was little or no effect on cobra venom as this type of poison is primarily a neurotoxin.

With modern advances in food preparation and refrigeration, incidences of **botulism** (a deadly form of food poisoning) have lessened considerably. However, we must never forget either just how easily it can be contracted or just how serious it can be considering the extreme toxicity of this form of bacteria. UVBI has proved to be extremely effective in the treatment of botulism, with the toxic symptoms being relieved within 24 to 72 hours of irradiation. Toxin neutralization is quite a remarkable and mysterious action of UV light therapy, but this detoxifying action is what makes UV light extremely useful in the preparation of various vaccines. UV irradiation will inactivate the toxic properties of bacteria and viruses without destroying the ability of the toxins to act as vaccine protective agents in humans and animals. This happens because the bacteria killed by UV light lose their toxicity but completely retain their antigenic (ability to make the body produce antibodies) and immunizing characteristics.

Because of the thousands of successful treatments carried out abroad, we have imported the UVBI equipment (technically known as the **OXYSAN** device) in order to commence the therapies in India. The device is basically a closed unit containing ultraviolet lamps through which a transparent, special quartz tube carries the blood for irradiation. Recently, we have completely cured two chronic cases. In the first case the patient had lapsed into a coma due to a blood clot in the brain, was not responding to any medication whatsoever and was on the verge of lapsing into a "brain dead" state. Five irradiation treatments later, this patient had fully regained his senses and was totally aware of his surroundings and what he was going through. He continues with the treatment. In the second case the patient had bronchitis coupled with an extremely sore throat for three years and had not responded to any allopathic medication (antibiotics, etc.) or treatment. After four irradiation sessions his sinuses have completely cleared, the throat soreness has vanished and the bronchitis is well on the way to being totally regressed.

So, What Is There to Lose?

Absolutely nothing! Not even time! UVBI is an efficacious treatment for almost every conceivable type of disease with absolutely ***no*** debilitating side effects. In addition, it is as safe as it is effective. There is no discomfort to patients undergoing treatment, the protocol takes no more than 45 minutes to complete and the beneficial effects sometimes become evident almost immediately. There are literally thousands of applications of the photoluminescence therapy—***ALL*** of them positive—but we're not going into specific details because there's no real need. Suffice it to say that UVBI therapy is well known, scientifically proven, totally effective, easy to carry out, relatively inexpensive and ***often resorted to even by orthodox physicians.***

CHAPTER 9

HYDROGEN PEROXIDE THERAPY: NATURE'S OXYGENATION PROCESS

Peroxide Good? No Way!

What with an abundance of free radicals and all, peroxides were always understood as being ***bad*** for the body. Now we have to believe the contradiction that **hydrogen peroxide** is actually ***good*** for us? A difficult pill to swallow no doubt, but there ***also happens to exist 176 years of medical literature on hydrogen peroxide*** which has either not been properly studied or has been purposely shoved into the background by the excitement generated at the advent of antibiotics. It is only now, antibiotics having finally been proved to have fallible limitations, that these valuable tomes on hydrogen peroxide have been dusted off and reopened. Wonder of wonders, we suddenly possess 'new' knowledge about an old 'something' that was ***brushed off as useless***. Hydrogen peroxide, also known as $\mathbf{H_2O_2}$, which should properly be called hydrogen dioxide (confusing?), is a colourless and odorless liquid that is soluble in water at all concentrations. It is normally used as (1) a bleaching agent, (2) an antiseptic and disinfectant, (3) an oxidizing agent (adds oxygen to other substances) and (4) an oxidizer in rocket motors for small rockets. Also, hydrogen peroxide occurs only in traces in nature, mostly in rain and snow. Basically, H_2O_2 puts extra oxygen into the blood and, when taken intravenously, rapidly relieves allergic reactions, such as influenza symptoms and acute viral infections, due to the oxidation of various foreign substances in the blood.

History of Intravenous Peroxide

In the early 1900s there was a world-wide epidemic of influenza/pneumonia wherein there was an 84% death rate. In 1920, in India, T. H. Oliver reported that he had taken 25 such critically ill Indian troops and injected them with hydrogen peroxide resulting in a 52% recovery rate. It was Dr. Charles Farr, of Oklahoma City, Oklahoma, USA, who eventually proved that the metabolic rate increased in those receiving a peroxide IV and that there was also an increased oxygen uptake. Medical literature has cited numerous situations in which arterially injected peroxide was of great clinical value. The efficacy of peroxide in a particular situation is indeed dependent on the route of administration. What was the mechanism of these remarkable recoveries from a terminal condition? If the oxygen is stimulating the good cells then it stands to reason that the bugs causing the malady should also be stimulated. But that does not happen. Actually, the toxins formed by the bacteria or virus get ***oxidized*** by the oxygen. It was always assumed that hydrogen peroxide was toxic to cells; now we know that it is an ***essential metabolite***. That means that far from being toxic, ***it is essential to life's process***.

What Are Its Properties Like?

Peroxide will react differently on humans and animals. This is mainly because of the content of ***catalase enzyme***. Animals like dogs, cats or chickens have very low catalase levels and can be killed by H_2O_2 administrations. Man, on the other hand, has an abundance of catalase in both the plasma and red cells and so has much higher tolerance levels. Peroxide is a well known but seemingly poorly understood compound which, as already stated, is odorless, boils at 152 degrees Celsius, decomposes at 84 degrees Celsius and is soluble in water. When water plus peroxide is distilled, the condensate will still contain some peroxide. The enzyme catalase is responsible for the conversion of H_2O_2 into water and oxygen. A very small amount of catalase can convert a huge amount of peroxide into water. It has been said that the peroxide in rain and snow keeps the surface of the earth from becoming covered with a putrid overgrowth of undesirable germs. A peroxide treatment causes oxygen to be dissolved in the serum and in the tissues. Oxygen is provided to the tissues, directly by serum, independent of that being delivered by the red blood cells. This oxygenation effect is relatively short-lived.

The release of oxygen from the peroxide into the body is not the primary benefit of peroxide therapy. The H_2O_2 initially reacts with catalase in the plasma and the white blood cells. It then penetrates the cell membrane of red blood cells where it reacts with catalase ***within*** the cell resulting in the release of additional oxygen. Some studies have been carried out on the veracity of hyperbaric oxygen—administering oxygen under high pressure in a high pressure tank—and intravenous H_2O_2 insofar as comparing the increased levels of oxygen content in the tissues is concerned. It has been found that, though the effects of hydrogen peroxide only start to register about 40 to 45 minutes after commencing the injection, the increase in tissue oxygen levels with both treatments is ***the same***. This is a very important finding because ***(a) hyperbaric oxygen therapy is expensive, (b) poses some risks, (c) is rather cumbersome and (d) is generally not available***. Peroxide, on the other hand, is relatively cheap, completely safe, readily available and extremely effective.

Blood samples taken during and after H_2O_2 infusions will show a colour change consistent with an increase in oxygen content of the blood. After an hour of infusion, there will be a 2 to 10 percent decrease in many of the blood's constituents such as sodium, potassium, chloride, phosphorus, etc. However, within 24 hours the levels of all these constituents will have returned to normal. The clinical benefits (of tissue fluid saturation by oxygen produced by H_2O_2) are of secondary importance because only small amounts of peroxide are used in the treatment and, therefore, very little oxygen is actually produced. What is important is that H_2O_2 is a powerful oxidizer and will oxidize both toxic and non-toxic substances alike. This action is completely separate from its role of oxygen contributor. The biologic effects could also be termed as '*oxidative detoxification*', which may include the oxidation of lipid material in the vessel wall to reverse atherosclerosis. Peroxide ***can*** be visualized as the ammunition of the body's killer cells, or 'bacterial assassins', called **polymorphonuclear leukocytes** (PMNs). These leukocytes engulf the bacteria and then kill them with the 'respiratory burst'. The cell combines oxygen and water making H_2O_2 (that's the respiratory burst) which in turn annihilates the bacteria. If the white cells didn't produce H_2O_2, then this respiratory burst would not be possible.

Hydrogen peroxide, therefore, besides truly being the wonder molecule, is one of life's most important bodyguards. The cells in the body that fight infection (called **granulocytes**) produce H_2O_2 as a first line of defence against ***every single type of invading organism***, be it parasite, virus, bacteria or yeast. In its importance to life on earth, ***no other chemical compound comes even close to H_2O_2***. It is involved in all of life's processes such as the metabolism of proteins, carbohydrates, fat, vitamins and minerals. Immunity and anything else involving life's functions require the presence of this amazing molecule.

The Actions of Hydrogen Peroxide

Hydrogen peroxide is a necessary ingredient in or a by-product of a number of chemical reactions which take place in our bodies. Let's take a look at a few of them:

PHAGOCYTOSIS. Few doctors realize that when a white blood cell engulfs a "germ" for phagocytosis that, under the influence of interferon, hydrogen peroxide is produced and secreted into the vacuole containing the undesired organism. If this organism contains no catalase, it is very likely to be destroyed (why give interferon when you can give peroxide?)

MEMBRANE INTEGRITY. Hydrogen peroxide is generated by most cellular membranes including mitochondrial and nuclear membranes. Peroxide maintains membrane integrity, and participates in the transport of some substances across membranes.

THYROGLOBULIN AND THYROXINE. Iodine requires peroxide in order to produce thyroglobulin and thyroxine.

EFFECT ON IRON AND LIPID PEROXIDATION. Lipid peroxidation occurs when the ratio of Fe-3 (ferric) to Fe-2 (ferrous) in the body is between 1.1 and 8.1. Natural levels of peroxide in the body tend to keep the iron ratios in this range and favor lipid peroxidation (that's why anti-oxidants are recommended by doctors). Fortunately, intravenous additions of peroxide cause excess ferrous iron to be oxidized into the ferric state. This keeps the ratio well above 8.1 and reduces lipid peroxidation. (NOTE: EDTA chelation therapy works similarly; by chelating away ferrous iron, the ratio of irons again is thrown to greater than 8.1 and lipid peroxidation is reduced). There is no study indicating that peroxide "initiates or supports" microsomal lipid peroxidation.

INSULIN-LIKE ACTION. Peroxide acts like insulin by affecting the RBC membrane to transport glucose. Blood sugar has been known to normalize during peroxide therapy.

PROGESTERONE PRODUCTION. Progesterone production in the uterus requires peroxide, which is produced under the stimulation of estrogen.

CONVERSION OF ARRHYTHMIAS. Peroxide was used successfully to convert ventricular fibrillation.

REMOVES PLAQUE. Termed also as 'oxidative detoxification', oxidation may remove lipid deposits from artery walls. Numerous studies suggest this benefit; however several studies involving 'intra-arterial' infusions, ***not intravenous***, have shown that intra-arterial injections cause a rise in the intra-arterial cholesterol and triglycerides followed by evidence of reduction of plaque and improved artery elasticity.

CORONARY ISCHEMIA. Peroxide has a protective effect on ischemic heart muscle.

CONTROL OF BODY TEMPERATURE. Peroxide participates in the control of body temperature through an effect upon the mitochondria.

"DEBRIDES" THE LUNGS. Presumed to be the result of high concentration of oxygen and peroxide at the alveolar surface, foreign debris—even of many decades duration—is oxidized and separated from the alveolar surface.

KILLS MANY VIRUSES, BACTERIA, YEAST, PROTOZOA, FUNGI, AND PARASITES. There is a big difference between lists of organisms killed in the test-tube and the practical application of intravenous peroxide. But there is no doubt that a peroxide IV kills or destroys a great many viruses, bacteria, fungi and parasites clinically.

DESTROYS COMPLEX MOLECULES. Although experimental evidence is lacking, complex and perhaps synthetic molecules for which the body has no method of detoxification or elimination, seem to be neutralized, oxidized, disassembled or otherwise eliminated. There seems no doubt but that, clinically, environmental sufferers and allergy sufferers improve with peroxide infusions.

REDUCES PAIN. There is no agreed explanation for the pain relief which may follow intravenous peroxide infusions. However, the theory is that most chronic pain results from anoxia (abnormally low oxygen) on a tissue level. It is believed that it is the oxygenation effects that are pain relieving. It has been noticed that when peroxide helps relieve pain that the effects remain for many hours, gradually wearing off as the extra oxygen provided by the treatment dissipates. When there is lasting pain relief, this is believed to occur from other mechanisms of action such as the killing of a virus, reduction of inflammation, etc.

Any Other Actions of Hydrogen Peroxide?

There are plenty of other beneficial actions precipitated by H_2O_2. Some of the more important ones are as follows:

- Cholesterol and triglycerides decrease immediately after intravenous infusions.
- Repeated infusions of peroxide do not cause bone marrow depression or anemia.
- Glucose may rise slightly due to the carrier solution, not from the peroxide.
- There may be reduction of RBC's and WBC's immediately after the treatment because old deteriorating cells have little catalase and are destroyed, leaving behind only good quality young cells. This temporary decrease has never been found to be deleterious.
- A blood chemistry profile is little changed 24 hours after a peroxide infusion; however there is a 2-10% decrease in many constituents within an hour of infusion.
- Peroxide has a role in prostaglandin synthesis.
- ATP production is indirectly stimulated.
- Coronary, pulmonary, and peripheral vessels are dilated.
- Heart rate and vascular resistance are decreased. Stroke volume and cardiac output are increased.
- Monocytes and T helper cells are stimulated.

Therapeutic Uses of Intravenous Peroxide

There are plenty of other diseases for which the use of H_2O_2 has yielded beneficial results. Some of the more important ones are as follows:

- Acute and chronic viral infections
- Alzheimer's disease
- Arrhythmias
- Asthma
- Cardioconversion
- Cardiovascular disease
- Cerebral vascular disease
- Chronic candidiasis (yeast infection)
- Chronic obstructive pulmonary disease
- Chronic pain syndromes (multiple etiologies)
- Chronic recurrent Epstein-Barr infection

- Chronic unresponsive bacterial infection
- Cluster headaches
- Coronary spasm (angina)
- Diabetes type II
- Emphysema
- Environmental allergy reactors (Universal)
- Herpes simplex
- Herpes zoster
- HIV infections
- Influenza
- Metastatic carcinoma
- Migraine headaches
- Multiple sclerosis
- Parasitic infections
- Parkinsonism peripheral vascular disease
- Rheumatoid arthritis systemic
- Temporal arthritis
- Vascular headaches

Does H_2O_2 Benefit the Heart In Any Way?

A lot of research has been done on what effects hydrogen peroxide has on the blood vessels. It is known to clear up arteriosclerosis and is positively seen to remove plaque buildup when injected directly into the arteries. The most amazing fact is that these improvements are ***not temporary***! Autopsies done on patients, who had undergone H_2O_2 therapy and died many years later, have proved beyond any doubt that the beneficial effects can, with complete confidence, be labeled "*permanent*". What is even more mind-boggling are the findings of research conducted with H_2O_2 and cardiac resuscitation.

Due to ***ventricular fibrillation***, a deadly and ominous event wherein the heart muscle literally goes crazy with a rapid rhythm that is totally chaotic in nature, victims of heart attacks often die within hours of the infarction having struck. This happens because of hypoxia–or the heart's response to the lack of oxygen. If this dangerous situation of the "heart going wild" can be controlled, then the patient's survival chances increase by leaps and bounds. There ***do*** exist some emergency measures—like electric shock or defibrillation and injecting certain cardiac drugs into the veins—to calm the heart down, but these have proved only partially successful. What must be remembered is that the heart is responding to a lack of oxygen, so any treatment—***other than supersaturating the blood directly with oxygen***—will only have temporary benefits.

Hydrogen peroxide has an energizing or **inotropic** effect on the heart's muscle, causing it to beat with higher vigor and efficiency besides dramatically helping the particular hearts that start to exhibit the inability to efficiently pump blood. "High output heart failure" leads to death because there is a back-up of fluid in the lungs with consequent drowning. Emergency applications of hydrogen peroxide (1) slow down the unhealthy, rapid heartbeat, (2) appreciably reduce blood pressure to normal levels, (3) dramatically improve the condition of **myocardial asemia** or lack of oxygen to the heart muscle and (4) completely relieve ventricular fibrillation.

Some of the More Beneficial "Combinations"

H_2O_2 and Cancer. Radiation therapy for cancer works in two ways on the human body. In the first instance, radiation ***does*** shrink the cancerous tumor, thereby improving the health of the patient. In the second instance, it also tends to shrink the body's immune system, thereby putting the patient's life at risk. There is also a direct relationship between the amount of oxygen present in the cancer mass (tumor) and the effectiveness of the radiation. The more oxygen that is present, then the radiation will be that much more lethal to the cancer. The easiest, fastest and safest way of ensuring this "extra oxygen" is by infusing hydrogen peroxide into the artery leading to the cancer. Then radiation treatment is administered and ***no cancer has any chance of proliferating after that!***

H_2O_2 and Bronchiectasis. This malady mainly involves the formation of pus pockets within the lung. Patients constantly cough foul-smelling phlegm, are always short of breath (often with a blue tinge in the face) and are invariably in a very weak physical state due to the lack of proper breathing. Hydrogen peroxide brings about total normalcy within a very short time.

H_2O_2 and Temporal Arteritis. This malady is characterized by severe pain and tenderness to the touch of the main superficial artery of the temple. If diagnosis is delayed in such a case, it can lead to blindness. Since it is also a proven process in the treatment of pneumonia and asthma, intravenously applied H_2O_2 clears up this condition completely as well.

H_2O_2 and Shingles. This disease involves inflammation of the nerve endings that is caused by the chicken pox virus. Painful and ugly blisters appear on the skin along the distribution of a nerve from the spine. Peroxide infusions remove the pain and dry up the blisters in most cases; however, there have been instances where peroxide treatment has not worked at all on this malady.

H_2O_2 and Chronic Obstructive Pulmonary Disease (COPD). This malady had, thus far, not been curable. It involves the constant presence of mucous and pus that threaten to block the respiratory passages. Scarring, as a result of the chronic inflammation and spasm of the bronchial passages, adds to the problem. Treatment of such cases used to be unending—until H_2O_2 was tried out as a possible alternative. Intravenous hydrogen peroxide will cause the mucous to bubble up and leave the lungs through a coughing fit. It was noticed during the treatment that the coughing and mucous production could be turned on and off by simply turning the H_2O_2 drip on and off!

H_2O_2 and the Yeast Syndrome. The results of treating yeast, or candidiasis, with H_2O_2 truly defy reason! In a lot of cases the known therapies for yeast—such as nystatin, acidophilus, caprylic acid, herbs, yeast extract injections for desensitization, ketoconazole (nizoral) and even homeopathics— have proved to work in fits and starts or yield no results at all. At most, all these therapies have yielded only temporary relief. Two sessions of H_2O_2 therapy give immediate, positive results and a total of between eight to ten sessions is enough to send the malady ***into complete recession***!

H_2O_2 and the Flu Syndrome. Flu, and other acute respiratory infections, is a condition for which hydrogen peroxide will probably find its greatest use. Also known as **viremia,** its classic symptoms are fever, chills, sore throat, cough and aching in the bones. The market is flooded with all kinds of flu suppressants that come in fancy packaging and have even fancier names—not to mention the fancy prices! None of these medicines comes even close to the results that are obtained by H_2O_2

therapy. A documented report reveals that in one case a 67-year-old man was put on peroxide drip. His fever dropped after one session and during the course of the second, ***even before the infusion was finished***, his temperature had become absolutely normal and all the symptoms ***completely vanished***! Such amazing results are almost boring in the monotony of repeating almost every case of flu that has been treated via H_2O_2—first day sick; second day 90 percent better; third day completely normal!

H_2O_2 and the Teeth & Gums. After reading this, it's most likely that you'll throw out your toothpaste and toothbrush! In the everlasting struggle to save one's teeth, there is nothing more useless than toothpaste. The dangers associated with fluoride use are well known and it is very difficult to find a toothpaste (in case you absolutely ***HAVE*** to use one) that doesn't contain fluoride (which is actually an enzymatic poison) or aluminum and silicon. The idea that the use of toothpaste will prevent tooth decay is a total fallacy that defies even the known facts of microbiology—let alone common sense! Even the assumption that food particles stuck between the teeth cause tooth decay and cavities has never been proved to be true. In fact, available evidence shows just the opposite—cavities are not the result of rotten food, ***they result from a rotten diet***. Brushing your teeth with H_2O_2 mixed with a little water and baking soda (make a thick solution not a paste) will have far better—and longer lasting—effects than any toothpaste. Gargling with three percent peroxide and swishing this around the mouth morning and night will even obviate the need for regular brushing. Hydrogen peroxide clears up every known mouth/tooth/gum infection that is known to medical science, and probably, those that are unknown also! So why not pack up your dentist and throw him away? The availability of hydrogen peroxide far exceeds the availability of dentists anyway!

CHAPTER 10

DISEASES THAT ARE THE SCOURGE OF MANKIND

To Know Is to Appreciate!

Though we have already spoken about heart and other diseases in the earlier chapters, the mentions are specific to the various alternative treatments available and how the cures happen. This chapter is meant to single out and explain, in detail and as simply as possible, the main diseases that we in India (*along with the rest of the modern world!-ed.*) are commonly susceptible to, how and why they occur, what precautions we must exercise and how these diseases may be overcome and avoided. This knowledge is absolutely essential in the light of recent medical research and findings. This chapter will be laid out in the form of a sort of "ready reckoner" so that you are able to ***easily recognise symptoms***, ***make a fair assessment of the severity*** and ***initiate timely corrective action***. We will, naturally, concentrate more on the heart and the blood—since both of these, somehow or other, always end up as the prime targets of various "ill effects", no matter where in the human body a particular malady may have first originated. Other diseases to be addressed are cancer, diabetes, and renal (kidney) disease.

Coronary artery disease is rapidly emerging as the undisputed, Number 1 killer on the disease front in India. Current estimates indicate that over 50 million Indians have contracted this malady in various degrees of severity. What is more alarming is that there are cases where the disease has been detected ***in new-born babies***. The World Bank has estimated that, taking into account the galloping progress of urbanization, by the year 2010 approximately 100 million Indians will have fallen victim to coronary artery disease in some form or other. In the beginning of the book, we mentioned that most of the current maladies are diseases of modernization and we reiterate that fact here. If it appears that our intention is to alarm you, you are *absolutely right!* ***That* is *our intention,*** mainly because ***forewarned*** means ***forearmed***. While going through these paragraphs, you are advised to refer back to chapters 2, 3 and 4 for a clearer insight into the problems caused by heart disease.

How Your Heart Affects You

Everyone knows that the heart is a vital organ and that no one can live without it. However, when you get right down to the nitty-gritty, the heart is nothing but a pump. An important and complex one, no doubt – but still just a pump! And, as with all other pumps, it is highly prone to becoming clogged, breaking down and needing repair. This is why it is critical that we know how the heart works. A little knowledge about the heart—*and what is good or bad for it*—will go a long way toward significantly reducing the risk of contracting heart disease, which is the leading cause of death in the Indian sub-continent. Did you know that we Indians are four times more prone to heart disease than Chinese, Japanese, Caucasians and Hispanics? Hardly anything to be proud of—right? On the flip side there is *some* good news in that the death rates from heart disease have been steadily decreasing. This is far from comforting considering the fact that in another ten years there will be 100 million of us walking around with coronary artery disease! Unfortunately, heart disease is still the main cause of sudden death and many people die before even reaching the hospital.

The heart holds a special place in our collective psyche besides, of course, it also being synonymous with love. But it also has many other associations. What other word has as many meanings as the word "heart"? Here are just a few examples:

- **have a heart** — *be merciful*
- **change of heart** — *change your mind*
- **know something by heart** — *memorize something*
- **broken heart** — *to lose love*
- **heartfelt** — *deeply felt*
- **have your heart in the right place** — *to be kind*
- **cry your heart out** — *to grieve*
- **have a heavy heart** — *sadness*
- **have your heart set on** — *to want something badly*

There certainly is NO other bodily organ that elicits this kind of response. When was the last time that you had a "heavy pancreas" or had a "change of liver"? In this write-up we will look at this most important and definitely most talked-about organ so that ***you understand exactly what makes your heart tick and how it affects you!***

Review of the Heart and Blood Circulation

Before we proceed with our discussion of heart disease and the associated symptoms and diagnosis write-ups, let's revisit and refresh our newly-acquired knowledge (from Chapters 2 and 3) about the 'Team Leader'—**our heart**.

The heart consists of two pumps side-by-side. The right and left sides of the heart have totally separate functions. The *right side* of the heart collects oxygen-depleted blood from the body and pumps it to the lungs where the blood is re-oxygenated and carbon dioxide is released. The *left side*, thereafter, collects this oxygen-rich blood from the lungs and further pumps it to the rest of the body so that every single cell is supplied with the much-needed oxygen to ensure proper function.

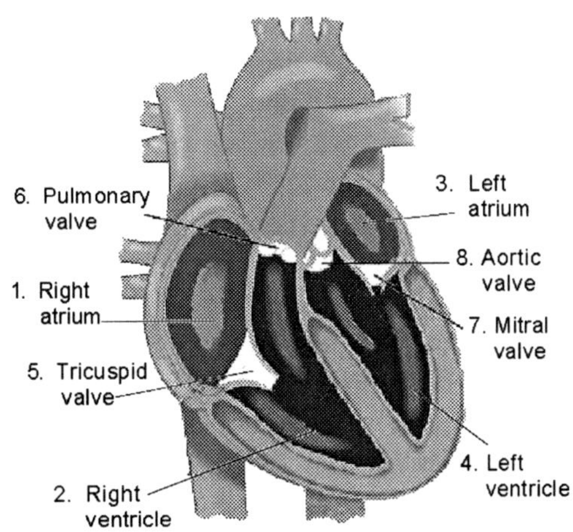

Figure 9 - Chambers & Valves of the Heart

As we have learned, the heart has four chambers (see Figure 9). The atria, or two top chambers, are for blood-collecting while the ventricles, or the two bottom ones, are the pumping chambers. The four chambers are [1]the **Right Atrium (RA)**, [2]the **Right Ventricle (RV)**, [3]the **Left Atrium (LA)** and [4]the **Left Ventricle (LV)**. Each chamber has a non-return or one-way valve at its exit that prevents the blood from flowing backwards.

What keeps the heart beating, as discussed previously in the BOOGIE TO THE BEAT section of Chapter 3, are the precisely timed electrical signals that originate within the heart and usually start in the sinoatrial (sinus node) at the top of the right atrium. This is also known as the heart's pacemaker. Nerve fibres conduct the electrical impulses from the sinus node to the atrioventricular node located between the atrium and ventricle chambers. Among its many functions, the atrioventricular node, in concert with the sinus node, transmits the exquisitely timed signals that control heart chamber contraction and expansion that pumps the blood through our body. All this activity produces electrical waves that are measurable and are used in diagnosis using a device called an **Electrocardiogram** (EKG). We'll discuss this in more detail later in this chapter.

What causes the sound that your heart makes? If you listen to your own heart through a stethoscope, what you will hear is commonly described as a "lub-dub, lub-dub". The first heart sound (lub) is caused by the acceleration and deceleration of the blood and vibration of the heart at the time of the closure of the **tricuspid** and **mitral** valves. The second sound (dub) is caused by similar acceleration and deceleration of blood and heart vibration at the time the **pulmonary** and **aortic** valves close.

Recalling from the FLOW OF BLOOD and the HEARTS BLOOD SUPPLY sections of Chapter 3, there are two primary blood circulation subsystems: body systemic and pulmonary. Plus there is the coronary subsystem that supplies the heart. The easiest way of reviewing the pulmonary and body systemic subsystems is by the adjacent flow chart describing blood circulation.

Oxygen-depleted blood from the body returns to

⇩

the *Superior* and *Inferior Vena Cava* and then

⇩

to the *Right Atrium* and

⇩

through the *Tricuspid Valve*

⇩

to the *Right Ventricle* for a further push

⇩

through the *Pulmonary Valve*

⇩

to the *Pulmonary Artery* and on

⇩

to the *Lungs* where oxygenation takes place, then flows out

⇩

through the *Pulmonary Veins*

⇩

to the *Left Atrium* and

⇩

through the *Mitral Valve*

⇩

to the *Left Ventricle* for a high pressure push

⇩

through the *Aortic Valve*

⇩

to the *Aorta* for distribution

⇩

to the rest of the body

⇩

go to the top and repeat the cycle

That special subsystem for the heart, the coronary system, provides the oxygen and nutrients required by our continuously working heart. Being a muscular pump responsible for moving massive amounts of blood every minute during peak exercise, the heart has equally urgent and huge needs for blood, oxygen and nutrients. As the blood is pumped out of the heart and into the aorta, a portion is routed back to the heart via the coronary arteries, three separate vessels which cross the surface of the heart, dividing, branching and ultimately penetrating the heart to supply it with oxygen and nutrients. As the workload increases with exertion or other stress, so does the heart's demand for oxygen. A healthy heart can cope with this extra demand, but a weakened one simply cannot. We'll find out later in this chapter how dependent we really are on the coronary subsystem.

Heart Disease - The Folly of Mankind

Despite the massive and continuing world-wide publicity of the various risk factors involved, heart disease, unfortunately, continues to be the world's number one killer and the malady appears to thrive even in the face of all odds! It has not only kept pace with modernization, but is currently almost at the stage of 'overtaking' industrialization! No one can pinpoint exactly why this is happening, but the best answer would probably be 'human nature'. All of us who are educated, and even a large percentage of the so-called 'uneducated', know for a fact that one needs healthy practices in order to maintain good health. We know that we must ensure healthy diets (less fat and more fruits and vegetables) but we persist in wolfing down junk foods and every other kind of 'devastating' edible. We know that regular exercise is a must, but most of us are the original 'couch potatoes' who wouldn't recognise exercise even if it up and slapped us in the face! We know that smoking is a dangerous habit, yet most smokers continue to generate more atmospheric pollution than did the Krakatoa eruption! Unfortunately, *knowing* isn't the same as *getting things done* and, human nature being what it is, it's more than likely that even after the danger has been recognised, one is just likely to turn his back and hope that the peril *will pass him by*—something like "what the eyes don't see, the heart won't grieve over". It really makes one wonder if the human race isn't actually suicidal and hell-bent on self-destruction! For those of us who may suffer from a congenital (a condition that one is born with) or contracted infectious malady, the situation was probably not of our making and is out of our hands. ***But the same does not hold true for heart disease,***

and some other maladies, if these were* not *afflictions from birth. Most of the illnesses of the modern age are contracted either because of a lack of knowledge and recognition, or because of a "couldn't care less" attitude. We need to know about, understand and face these problems head on. It is well to remember that it is never too late to learn about what ails and assails our world on the health front and to initiate the positive and necessary corrective measures without a moment's delay.

Definition. Heart Disease may be defined as any condition that causes the heart to malfunction. Loosely applied, 'heart disease' is an all-purpose term for a whole kit and caboodle of debilitating conditions that may affect the heart. As already stated, some heart diseases may be congenital, but what we're more concerned with here is the far more common set of symptoms, dysfunctions and life-threatening aspects that come about largely due to the lifestyles that we lead. When *heart disease* is mentioned, what people are usually referring to is the coronary heart conditions that, more often than not, lead to heart attacks and angina which are ultimately caused either by arteriosclerosis (hardening of arteries) or atherosclerosis (artery blockage).

SYMPTOMS OF HEART DISEASE. There are various symptoms of heart disease that even a layman may easily recognise and distinguish. All it really requires is the acknowledgement that the body ***is***, in fact, sending out certain distress signals and needs to be attended to. Some of the very common symptoms are ***dyspnea*** (difficult or labored breathing), ***chest pain***, ***palpitations***, ***presyncope*** or ***syncope*** (loss of consciousness) and ***fatigue***. None of these is very specific and you, as the *sufferer*, need not bother about the interpretations which, in any case, would require diagnostic testing for the proper clinical picture to evolve. Your 'job' is to heed the signals, recognise, acknowledge and seek your doctor's opinion immediately. So, if you have been living a normal life and one day find that one (or all) of these symptoms is suddenly in evidence, it's time to sit up, ***pay attention*** and initiate the proper remedial action. A bit more discussion about each symptom follows:

- **Dyspnea**. Difficult or labored breathing is brought on and aggravated by exertion. It is usually the result of the *left atrium* and *pulmonary* (of the lungs) *venous pressures* being elevated and also because of **hypoxia** (lack of oxygen). High left atrium and pulmonary venous pressure are most commonly caused by left ventricular *systolic* and *diastolic* dysfunctions or a *valvular obstruction*. Hypoxia occurs either due to **pulmonary edema** (the arteries or veins getting swollen with watery fluids) or **intracardiac shunting** (movements of blood and fluids from chamber to chamber or within the heart structure). Dyspnea is also a common symptom of pulmonary (lung) disease, and it may be difficult for one to delineate an exact cause. The severity of dyspnea is dependent on the amount of activity which precipitates it. Shortness of breath may also be found in obese persons, in those having sedentary (sitting most of the time) lifestyles and in individuals suffering anxiety states, anemia and many other illnesses.

- **Orthorpnea** is dyspnea that is caused by pulmonary disease and obesity and is more specific for cardiac disease than is exertive dyspnea. However, it is not indicative of heart failure in either case.

- **Paroxysmal nocturnal dyspnea** is shortness of breath that occurs abruptly between 30 minutes to 2 hours of going to bed. Relief is obtained by either sitting up or standing up. This particular symptom is more specific for cardiac disease.

- **Chest Pain**. Chest pains may occur for various reasons that may, ***and also may not***, be related to heart disease. Sometimes the chest pains resulting from gastrointestinal disorders are very similar to those brought on by heart disease and, one is frequently mistaken for the other. The bottom line in such situations is that chest pain, regardless of the cause, is not to be taken lightly, and the reasons for the occurrence should be medically determined immediately. The most common cause of cardiac chest pain—***myocardial ischemia***—is a lack of blood supply to the heart muscle. This pain is known as **angina**. The salient sensations of ischemic pain may be recognised as a "dull ache", "pressure", "tightness in the chest", "squeezing", or "gas"—as *opposed* to being sharp or spasmodic. In fact, one would more likely describe it as a feeling of discomfort rather than actual 'pain'. Chest pains that last for 30 minutes or more and occur in a repeated and protracted manner are indicative of a heart attack. The pain will usually be accompanied by a sense of anxiety or uneasiness and will be located in the region of the breastbone, regardless of whether it eventually radiates to or is localized in the throat, lower jaw, shoulders, inner arms, upper abdomen or back. The pain is often triggered by exertion, cold weather, meals, stress or a combination of these factors. Relief is usually provided by a period of rest. Chest pain can also originate from other areas that are near the heart such as the ***pericardium*** (the sac around the heart), ***aorta***, ***lung***, ***pleura*** (lining of the lung), ***muscles and bones of the chest*** or the ***gastrointestinal tract*** (esophagus or stomach). Also, one may be afflicted with heart disease and ***display* NO *symptoms of chest pain whatsoever***.

- **Palpitations, Dizziness and Syncope**. If one experiences unsteady heartbeat (ventricular premature beats), the manifestation can be recognised as either "extra" or "skipped" heartbeats. When these become rapid, they may be felt as regular or irregular palpitations or "fluttering". Palpitations are an awareness of one's heartbeat and can occur in heart disease. This can cause a feeling that your heart is beating too fast or too slow, skipping beats, or beating irregularly. If this abnormal rhythm happens to be associated with a decline in atrium pressure or cardiac output, it will most probably interfere with the blood flow to the brain further causing dizziness, blurred vision or a loss of consciousness (syncope) and also bring about seizure-like movements. Syncope or passing out will occur if the heart is temporarily not pumping out enough blood. Nausea or vomiting and profuse sweating (***diaphoresis***) can sometimes accompany heart disease.

At this juncture you should not bother yourself with determining the causes for these symptoms—we'll leave the diagnostic work to the doctors, laboratories and clinics. Suffice it to say that once you are aware that some ailment ***is*** manifesting itself through such symptoms, you will then take the appropriate steps to have the matter investigated.

TYPES OF HEART DISEASE. There are a wide range of diseases of the heart. These include:

- **Angina.** Known medically as **angina pectoris,** this is characterized by a sense of discomfort or actual pain in the chest arising from the heart. Angina is *not* a disease, but is a *symptom* of disease. It is most commonly caused when the heart is not getting a sufficient supply of oxygen because the **coronary arteries** have become narrowed. Angina may also occur in normal or minimally narrowed coronary arteries when the oxygen demands of the heart exceed normal limits. The most common cause of this is **valvular heart disease**. Other potential causes include cardiac rhythm disturbances (arrhythmias), diseases of the muscular portion of the heart and non-cardiac conditions such as **hyperthyroidism** and **anemia.** Angina can present a variety of symptoms: it may feel like a mild discomfort near the centre of the chest or a burning sensation similar to heartburn and indigestion. It may also be described as a tightening, heaviness, cramping, squeezing or sharp pain or discomfort behind the breastbone. The sensations may also spread to the neck, jaw, throat, shoulder, upper back or arms. Angina does not always cause pain and for some people it may manifest itself as a suffocating sensation accompanied by sweating, breathlessness and weakness. The symptoms of angina are normally triggered during physical exertion when the heart is working 'overtime'. A heavy meal, intense emotion, and running or walking in very cold weather can also bring on these sensations. Usually the symptoms subside and disappear with rest after 15 to 20 minutes, but this does not hold true for everyone. The effects differ with different persons. There are also other conditions which tend to mimic angina and should not be mistaken for the real thing. Some common disorders with 'look-alike' symptoms are **gastroesophogeal reflux** or **common heartburn**, inflammation of or trauma to the chest wall or pain arising from stones and other abnormalities of the liver drainage system. ***While angina itself does not lead to permanent heart muscle damage***, it is indicative of the presence of heart disease which could eventually result in a fatal heart attack. The symptoms may occur in a number of patterns:

- **Chronic stable angina** causes predictable symptoms, typically after moderate or strenuous activity. Such a pattern is usually considered safe and may remain unchanged for years.

- **Unstable angina** occurs while at rest or with minimal or no provocation. The term also refers to a sudden worsening of the symptoms experienced in stable angina. These changes in symptoms may signal a heightened risk of heart attack. Unstable angina is a definite reason for hospitalization and urgent treatment.

- **Variant angina** is a unique and uncommon form that does not result from narrowed arteries or other underlying heart disease. It is caused by a temporary spasm of the muscle layer within the wall of the artery, causing narrowing. Attacks can be very painful, occurring at any time and with similar symptoms to other angina types. One major characteristic is that it occurs at rest and not in response to exertion. It may be dangerous since it could lead to both heart attack and potentially lethal heart-rhythm disturbances when it is severe or prolonged.

- ♦ **Microvascular angina** exhibits virtually all the symptoms of stable angina but occurs in patients with normal coronary arteries. This pain may be due to the inability of the coronary arteries' to dilate during exercise.

- **Heart Attack**. This normally occurs when one of the heart's arteries becomes blocked and the blood supply to the heart is shut off. Known as a **myocardial infarction**, the symptoms—similar to those of angina but far more severe and lasting for a longer period—would be a chest pain or discomfort, which may be light or heavy, accompanied by breathlessness, profuse sweating, dizziness, a thready and erratic pulse and sometimes nausea and vomiting. There is the mistaken contention that every time angina or chest pain occurs, it is the signal of a heart attack, while there are others who think that the heart stops beating with a heart attack. Both ideas are incorrect. Though heart attacks may eventually lead to heart stoppage, the correct term for a heart not beating is **cardiac arrest**.

- **Congestive Heart Failure (CHF)**. This condition occurs when the heart is unable to pump well enough to meet the body's demands (from tissues and other organs) thereby depriving the tissues of adequate oxygen and the much-needed nutrients carried by the blood. The leading causes of heart failure in people over forty are narrowed and hardened coronary arteries (the vessels that supply blood to the heart muscle), ***history of a previous heart attack***, blood clotting in the heart, uncontrolled high blood pressure and disease of the heart valves. In CHF, the weakened heart pumps less blood than usual—probably much less than the amount actually needed to push circulated blood back into the heart to repeat the journey. So blood stalls, or congests, in the body's tissues. This makes it even harder for the heart to push blood through the arteries on the next beat.

With congestion, legs and ankles swell. Fluid collects in the lungs and interferes with the ability to breathe, **especially when lying down**. Also, the kidneys have trouble disposing of sodium and water. The body retains water, increasing congestion and swelling. If this condition is left untreated, congestion will worsen and may prevent the heart from pumping enough blood to keep the person alive. Cases of CHF may be divided into four levels of increasing severity:

Class I. Physical activity remains unaffected. The patient does not experience any unusual fatigue, shortness of breath, palpitations or pain during normal activities.

Class II. The patient may experience mild fatigue, shortness of breath, palpitations or pain during normal activities and also have slight limitations on normal activities.

Class III. The patient experiences fatigue, shortness of breath, palpitations or pain during normal activities which must be severely curtailed.

Class IV. Unable to carry out any physical activity without discomfort; symptoms of heart failure are present even at rest, with increased discomfort upon any level of physical activity.

- **Valvular Heart Disease**. Diseases of the heart valves were at one time due to rheumatic heart disease but nowadays other causes are more common. The various diseases under this heading are:
    - Aortic regurgitation
    - Aortic stenosis
    - Endocarditis (infection of a heart valve).
    - Mitral regurgitation
    - Mitral stenosis
    - Tricuspid regurgitation
    - Tricuspid stenosis

Stenosis means the narrowing of a passageway; in this case the passages controlled by individual valves. ***Regurgitation*** means the inability of the valve to prevent a backflow of blood or, in terms of valvular performance, it also means insufficiency. All of the heart's valves are of the "non-return" variety that permit blood flow in only one direction. When they perform dysfunctionally, serious problems arise in the overall performance of the heart.

- **Ventricular Arrhythmia**. When asked to name their greatest health fear, many people generally list blindness, a fatal heart attack and cancer – in that order. Thanks to the widespread publicity given to atherosclerosis as ***the*** high-risk factor leading to myocardial infarctions, many such "cardiophobics" are of the firm opinion that their eventual demise will result from plaque-clogged arteries shutting off the blood supply to the heart. They couldn't be further from the truth! **Ventricular arrhythmias**, or failing and disjointed heart rhythms, are the most common and deadliest of all heart conditions. The heart is not just a mere vital muscular pump feeding life-sustaining blood throughout the vascular system, more importantly, it is more of a "hot-wired" organ dependent on steady and unfailing electrical impulses from its internal pacemaker in order to function efficiently. Cardiac rhythm disturbances—**arrhythmias**—are classified according to where they occur in the heart and their effect on heart rhythm. **Atrial arrhythmias** arise in the atria while **ventricular arrhythmias**, generally considered to be far more dangerous, begin in the ventricles. There are further classifications:

    - **Sinus arrhythmias.** Cyclic changes in the heart rate during breathing, common in children and often found in adults.
    - **Sinus tachycardia.** The sinus node sends out electrical signals faster than usual, speeding up the heart rate.
    - **Sick sinus syndrome.** The sinus node does not fire its signals properly, so the heart rate slows down. Sometimes the rate alternates between a slow (***bradycardia***) and fast (***tachycardia***) rate.
    - **Premature supraventricular contractions or premature atrial contractions (PAC).** A series of early beats in the atria speed up the heart rate. In paroxysmal tachycardia, repeated periods of very fast heartbeats begin and end suddenly.
    - **Atrial flutter.** Rapidly fired signals cause the muscles in the atria to contract quickly, leading to a very fast steady heartbeat.

- **Atrial fibrillation.** Electrical signals in the atria are fired in a very fast and uncontrolled manner. Electrical signals arrive in the ventricles in a completely irregular fashion resulting in a heart beat that is quite erratic.
- **Wolff-Parkinson-White syndrome.** Abnormal pathways between the atria and ventricles cause the electrical signal to arrive at the ventricles too soon and it is transmitted back to the atria. Very fast heart rates may develop as the electrical signal ricochets between the atria and the ventricles.
- **Premature ventricular complexes (PVC).** An electrical signal from the ventricles causes an early heartbeat that generally goes unnoticed. The heart then seems to pause, in a regular fashion, until the next beat of the ventricle occurs.
- **Ventricular tachycardia.** The heart beats fast due to electrical signals arising from the ventricles (rather than from the atria).
- **Ventricular fibrillation.** Electrical signals in the ventricles, fired in a very fast and uncontrolled manner, cause the heart to quiver rather than to beat and pump blood.

There's very little that we *do* know—and a whole lot more that we *don't*—about the underlying causes of cardiac arrhythmias, though physicians have known about the electrical malfunctions of the heart since 1774.

Something You're Familiar With, But Know Little About!

There are two very commonly discussed maladies that are talked about quite freely, but without any real understanding of the various complexities involved. In fact, these occur so frequently and are so commonplace, that they are generally taken for granted and, with 'jet set' medicines affording immediate relief, are normally considered everyday occurrences that shouldn't be bothered about. We shall give you details to prove that neither are these as *common* as one would like to think, nor are they something to be shrugged off. If not tended to in time and in depth, these could ultimately imperil your very life! We're talking here about ***angina*** and ***blood pressure***.

Coronary Artery Disease. *Coronary artery disease* (CAD), *coronary heart disease (*CHD), *ischemic heart disease* (IHD) and *arteriosclerotic cardiovascular disease* (ASCVD) are the different nomenclatures for the same malady. The chest pain resulting from this malady is what we call **angina,** which was extensively discussed in the preceding section. CAD is caused by atherosclerosis, which entails the build up of fatty deposits (atheroma) in the coronary arteries. Since the coronary arteries supply blood to the heart muscle, a blockage in any one of these means a decreased supply to that muscle. Under conditions of exertion the heart is laboring harder and hence requires a greater volume and supply of oxygenated blood to enhance its strength. Any blockages will prevent the blood flow from increasing as needed during exertion, resulting in the heart muscle suffering from a lack of oxygen and pain—*angina*—being induced. The heart muscle is said to be ischemic when it lacks sufficient oxygen. Death of a cell is known as an infarction. Because a heart attack indicates heart muscle (myocardium) cell death, it is also termed *Myocardial Infarction.* The process that results in CAD, angina and finally a heart attack is called *atherosclerosis* which is a type of *arteriosclerosis*—the more general term for hardening of the arteries. If you flip back to THE CALCIUM-CHOLESTEROL COMBINE section in Chapter 4 for a detailed description of both these conditions, you will find that atherosclerosis causes the build up of fatty substances along the inner lining of the arteries. These substances, also known as atheromas and plaques, are responsible for lessening the flow of blood by causing blockages.

Blood Pressure. Blood pressure is an important diagnostic index, especially of circulatory function. Because the heart can pump into the large arteries a greater volume of blood than that which can be absorbed by the tiny arterioles and capillaries, a resulting back pressure is exerted against the arteries. Any condition that dilates or contracts the blood vessels or affects their elasticity or any disease of the heart that interferes with its pumping power, affects the blood pressure. When the blood pressure is measured, readings at the high point (the **systole** pressure) and at the low point (the **diastole** pressure) are determined and the results, in millimeters (mm) of mercury, are reported as a ratio expression of maximum over minimum, for example, 140/80. The interpretation of blood pressure readings by the National Institutes of Health are as follows:

SYSTOLIC PRESSURE (if diastolic pressure is below 90)

- Less than 140......Normal. Recheck within two years
- 140-159............Borderline isolated systolic hypertension. Confirm within two months.

- 160-200...............Isolated systolic hypertension. Confirm within two months.
- 200 or higher.......Refer for care within two weeks

DIASTOLIC PRESSURE

- Less than 85.........Normal. Recheck within one year
- 85-89....................High normal. Recheck within one year
- 90-104..................Mild high pressure or hypertension. Confirm within two months
- 105-114................Moderate hypertension. Refer for care within two weeks
- **115 or higher**......***Severe hypertension. Refer for immediate care.***

The blood pressure of an individual varies during the course of a day. It is lower when you are resting and tends to rise after exertion, especially after a sudden burst of activity. Many individuals find their blood pressure readings high in a doctor's clinic through the sheer anxiety of having the blood pressure taken! Therefore most prudent physicians will not prescribe pressure-lowering medications after the initial reading but will wait until a pattern of several high readings emerges. If one already has high blood pressure and is on medication, taking these should be regular as should the visits to your doctor.

When the blood pressure becomes abnormally high, the condition is known as **hypertension.** This happens when either arteriosclerosis or atherosclerosis is present. In both cases normal blood flow and circulation are hampered resulting in elevated blood pressure. High blood pressure can result in sweating, blurred vision, blinding headaches, rapid pulse and shortness of breath. This disease is brought on through cigarette smoking, stress, obesity, excessive use of stimulants, drug abuse and high sodium intake. Abnormally low blood pressure, or **hypotension**, is a result of infectious and wasting diseases, hemorrhage and collapse. Systolic pressure lower than 80 is usually associated with shock. As you progress through the following discussion of heart diseases, notice the number of times *hypertension* or high *diastolic* pressure is mentioned as a related problem or a symptom.

Pericardium Diseases. The thin, fibrous sac which encloses the heart, keeps it in the correct anatomic position and reduces contact between the heart and surrounding structure, is called the **pericardium**. It consists of two layers, an inner visceral layer that is attached to the epicardium and an outer housing. It is made up of fibrous tissue that will permit moderate changes in heart size. However these tissues cannot stretch fast enough to accommodate rapid heart dilation or an accumulation of fluids. The pericardium is prone to infection from diseases of adjacent tissues, systemic diseases, viral attacks, bacterial and fungal attacks, radiation and drug toxicity. Symptoms of this malady are fever, night sweats, chest pain, elevated jugular venous pressure and dyspnea. Some of the specific terms that you might hear in respect of these are:

- Acute pericarditis
- Constrictive pericarditis
- Inflammatory pericarditis
- Neoplastic pericarditis
- Pericardial effusion
- Postcardiotomy pericarditis
- Radiation pericarditis
- Tuberculous pericarditis
- Uremic pericarditis
- Viral pericarditis

Heart Muscle Diseases (myocardium). These will be primarily discussed under two headings, namely myocarditis (affecting the heart muscle directly) and cardiomyopathy (affecting the tissues of the heart muscle).

- **Myocarditis.** Acute myocarditis causes focal or diffuse inflammation of the myocardium. Most cases, generally known as *Infectious* or *Viral Myocarditis*, occur as a result of infection, rickettsial, spirochetal, fungal or parasitic agents. Immunologic disorders are also responsible. This form of myocarditis often follows an upper respiratory infection and may be accompanied by a pleuritic or non-specific chest pain or signs of heart failure. There would also be signs of the presence of arrhythmias and conduction abnormalities. The potential for the HIV virus to cause this form of myocarditis is also well known.

  The second form of myocarditis is *Drug-Induced & Toxic Myocarditis*. As the name implies, it is caused by drugs and toxins. A variety of medications, illicit drugs and toxic substances can produce acute or chronic myocardial injury. Protracted use and over-dosing will cause inflammation and necrosis (decay of body tissue) together with clinical heart failure and arrhythmias. Some drugs precipitate cardiac dysfunction while others can cause coronary artery spasm and myocardial infarction. Radiation can cause an acute inflammatory reaction as well as chronic fibrosis of the heart muscle, usually in conjunction with pericarditis.

- **Cardiomyopathies.** The cardiomyopathies (diseases of the heart muscle tissues) are a group of entities composed of diverse elements that affect the myocardium primarily and are ***not*** associated with the major causes of heart disease. While some have specific causes, many cases are **idiopathic** (a disease not preceded or occasioned by another or by any known cause). There are three main types:

  1. ***Primary Dilated Cardiomyopathy*** usually shows the signs and symptoms of congestive heart failure (most commonly dyspnea) and, occasionally, symptomatic ventricular arrhythmias are the first event. The heart tends to become enlarged with evidence of pulmonary congestion, left ventricular dilation, systolic dysfunction, elevated jugular venous pressure and severe bi-ventricular failure. Often no cause can be identified, but alcohol abuse and myocarditis are probably the most frequent reasons. Genetic diseases and glandular secretions flowing directly into the blood are also possible causes.

  2. ***Hypertrophic Cardiomyopathy*** is a hereditary syndrome and can also be brought on by chronic hypertension. The disease is characterized by a small, hypercontractile left ventricle. The presence of arrhythmias is an important problem with atrium fibrillation, a long-term consequence of chronically elevated left atrium pressures, being a poor prognostic sign. The most frequent symptoms are dyspnea and chest pain. Syncope may commonly occur, but this remains typically a post-exertional symptom. Ventricular arrhythmias are also common and could cause sudden death, especially in athletes after extraordinary exertion. Diastolic dysfunction is another primary symptom.

  3. ***Restrictive Cardiomyopathy*** is characterized by impaired diastolic filling with preserved contractile function. This is a very rare condition with the most common causes being amyloidosis, radiation, and myocardial fibrosis after open heart surgery and diabetes. The most frequent symptoms are dyspnea, fatigue and right-sided congestive heart failure.

Other symptoms may be the presence of conduction disturbances and ventricular hypertrophy. There will be elevated jugular venous pressure and high diastolic pressure as well.

Congenital Heart Disease. This particular malady basically comprises birth defects of the heart. The four most common lesions will be highlighted here.

1. ***Pulmonary Stenosis***. This is a condition where there is stenosis (narrowing) of the pulmonary valve or **infundibulum** (the valve being funnel-shaped or funneled). Because of this birth defect, the narrowing (or funneling) of the pulmonary valve increases the resistance to blood outflow, raises the right ventricular pressure, limits pulmonary blood flow and reduces cardiac output. Moderate to severe stenosis may cause dyspnea on exertion, chest pain and eventually right ventricular failure. Those with a mild stenosis may have a normal life expectancy. Victims with moderate stenosis may remain symptom-free in childhood and adolescence, but the symptoms increase as the patient gets older. Severe stenosis is associated with sudden death and can cause heart failure in the 20s and 30s. The condition can only be corrected either through balloon valvuloplasty or surgery.

2. ***Coarctation of the Aorta***. This malady consists of localized narrowing of the aortic arch a little away from the point of attachment of the left subclavian artery. A bicuspid (having two peaks) aortic valve is present in 25% of cases. Infants may have severe heart failure. However, if this does not occur in infancy, then children and adults usually remain asymptomatic till hypertension starts to show. The hypertension will further produce left ventricular failure or a cerebral hemorrhage. Hypertension will be present in the arms, but the pressure will be normal or low in the legs. This difference is exaggerated with exercise. Femoral (thigh area) pulses will be weak and delayed in comparison with the brachial (arm) pulse. Strong arterial pulsations will emanate in the neck and suprasternal notch. As already stated, cardiac failure is common in infancy and older untreated patients but is rare in late childhood and young

adulthood. Most of those with the adult form of coarctation, in whom the condition remains untreated, are likely to die before age 50 from complications of hypertension, rupture of the aorta or cerebral hemorrhage. The condition can be corrected through surgery by re-sectioning the coarcted site (actually a must for all patients up to the age of 20). However, the risks of surgery are many for patients of 40 years and above, besides the likelihood that they will remain hypertensive for years after the operation.

3. ***Patent Ductus Arteriosus***. In a fetus, the right and left sides of the heart actually operate in parallel to pump blood through the systemic organs and the placenta. Foetal blood that returns from the systemic organs and placenta fills **both** the right and left sides because of an opening in the intra-atrial septum (the 'wall' separating the left and right atria) called the ***foramen ovale***. The blood pumped by the foetal right heart does not enter the occluded pulmonary circulation but rather is diverted into the aorta through a vascular connection between the pulmonary artery and the aorta called the ***ductus arteriosus***. The foramen ovale is eventually closed permanently after birth by the growth of fibrous tissue. In the case of the ***uctus arteriosus***, through mechanisms that are not fully understood but are clearly linked to a rise in blood oxygen tension, this opening gradually constricts and completely closes over a period ranging from a few hours to a few days after birth. The problem arises when the embryonic ductus arteriosus fails to close normally, remains patent (open) and persists as a shunt connecting the left pulmonary artery and aorta. Prior to birth, the ductus is kept patent by the effect of circulating prostaglandins. If the defect is not closed after birth, blood flows continuously from the aorta through the ductus into the pulmonary artery in both systole and diastole increasing the work of the left ventricle. This will cause pulmonary hypertension or left ventricular failure to develop, normally during middle age. The heart size will be normal or slightly enlarged with low diastolic pressure. Large shunts cause a high mortality rate from cardiac failure early in life. Smaller shunts will permit a normal survival with congestive heart failure being the main complication. The ductus can be

closed through surgery without undue complications and should be carried out on children and adults who display mild symptoms. Cases showing chronic symptoms are often inoperable.

4. ***Atrial Septal Defect***. Similar to the foregoing malady, this particular disease involves the non-closure of the foetal ***foramen ovale*** after birth. If the connection between the left and right atria persists after birth, this will give rise to pumping and pressure problems as well as a disturbed blood flow. In all such cases, normally oxygenated blood from the left atrium passes into the right atrium, increasing right ventricular output and pulmonary blood flow. Most patients having small or medium unclosed openings generally remain asymptomatic. Those with large shunts (openings), more specifically after the age of 40, could develop exertional dyspnea, pulmonary hypertension, cardiac arrhythmias (especially atrial fibrillation) and heart failure. Patients with pulmonary hypertension or venous thrombosis (the coagulation of blood in a blood vessel or organ) are especially a cause for concern. The defect can be corrected through surgery, though those with small or medium shunts can live comfortably with the defect so long as they do not develop any symptoms.

Diagnosing Heart Disease – What's Going On Inside?

Modern medicine has a number of tools available to help determine what is happening to and inside your heart. This section describes the 'what' and 'how' of the customary diagnosis techniques.

Medical History and Physical Examination. Even in this technological age, wherein your physician has a wide assortment of tests available, the primary source of information about your heart comes through your medical history and a physical examination. Most people, especially those who enjoy generally good health, tend to ignore or minimize the importance of keeping a record of the various ailments (and the related treatment/medication) that one may have gone through at some time or other during life. Almost 100 % say that this is the doctor's job. That may be so in the technologically advanced countries where histories are computerised and stored from birth and are available at the click of a 'mouse', no matter where a particular patient may be. *(In*

America, wide spread computer-based record keeping is so recent that it doesn't cover patients from birth who are over 25 – 30 years of age! – ed.) We have no such system in India and it is generally the responsibility of the individual to collect, compile, store and maintain such information. A small piece of advice—in case YOU haven't initiated any such processes for yourself and your family, please do so immediately. An accurate medical record is of tremendous help to the diagnosing cardiologist—or any other physician, for that matter—for a precise and clear assessment of both the probable cause factors and the most appropriate treatment. It would also be expedient for you to update these records annually—even if you don't suffer any illnesses—with the basic parameters of height, weight, chest and waist measurements, pulse rates, blood pressure, eyesight, hearing etc. After one passes the age of 35, an annual EKG (discussed below) along with a complete blood profile wouldn't be such a bad idea either!

During a physical exam your doctor will normally first check your existing physical condition. Changes in height, weight, eyesight, hearing, pulse rates, blood pressure, reflexes, heart sounds, lung sounds etc., besides indicating the ageing factor, could point to the existence of a malady. Particularly for heart disease, your physician will carefully monitor your pulse through a stethoscope. For the heart, this instrument is placed over particular, pre-determined areas—particularly where each of the heart's valves sounds the loudest. If a murmur is heard loudest at a valve location, the chances are that it is ***that*** particular valve which is malfunctioning. The heartbeat can also tell a story of its own. Listening to the lungs can reveal the presence of fluid, which could result from heart disease.

Electrocardiogram. An Electrocardiogram (EKG—from the German spelling—or ECG) records the electrical activity of the heart from the surface of the chest using electrodes placed on each arm and leg and six locations on the chest. These electrodes are connected to an EKG machine. Twelve different tracings are produced by a

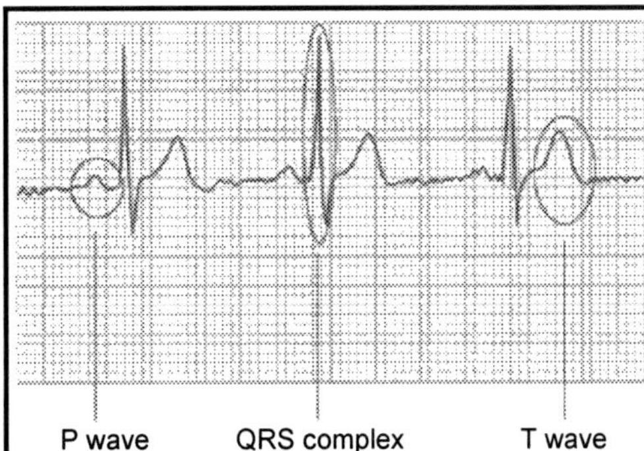

Figure 10 – Typical EKG Trace with Features Identified

12-lead EKG, and each tracing gives a view of the heart from a different angle. Depending upon which EKG lead is involved, your physician is able to determine in which part of the heart an attack is occurring. Thereafter, based on the knowledge of anatomy, he can also determine which of the arteries is blocked. The EKG gives the physician information about heart rate and rhythm, adequacy of blood supply to the heart, presence of a heart attack, enlargement of the heart, inflammation around the heart (**pericarditis**) and the effects of drugs and electrolytes on the heart. An example of three heartbeats from an EKG is shown on the adjacent illustration. Each part of the tracing has a lettered name that relates to a portion of the heart beat cycle as summarized by the following table:

P wave	Coincides with the spread of electrical activity over the atria and the beginning of its contraction.
QRS complex	Coincides with the spread of electrical activity over the ventricles and the beginning of their contraction.
T wave	Coincides with the recovery phase of the ventricles.

Echocardiogram. This is a painless procedure that uses an ultrasound beam to view the heart in motion. The procedure is similar to that used to monitor a fetus. An ultrasonic transducer, which looks like a microphone, transmits and receives the ultrasound waves. It is placed on the chest wall and moved around to view different heart structures. Ultrasound waves are reflected only when they reach the edge of two structures with different densities. The reflected waves produce a moving image of the edges of heart structures which are displayed on a screen and recorded on tape. The types of Echocardiograms are *M-mode, 2-D* and *doppler*. An M-mode is a one-dimensional view of a small section of the heart as it moves. The 2-D echocardiogram produces a moving two-dimensional slice of the heart. The Doppler ultrasound is used to evaluate the velocity and turbulence of blood flow in the heart, wall motion abnormalities that occur when the heart muscle is not receiving enough blood, and the presence of aneurysms, clots, tumors, and bacterial growths on the valves.

Echocardiograms can also evaluate:

- The presence of any abnormal fluid collection in the sac around the heart.

- The chamber size, thickness of the heart muscle wall and how well it is functioning.
- The function of the heart valves, whether they are obstructing blood flow or leaking.
- Any abnormal connections between chambers and vessels that may exist in congenital heart disease.

Chest X-ray. This is a useful tool in the evaluation of heart disease. It shows the size and shape of the heart, which may be enlarged in cases of congestive heart failure. Enlargement of the heart (**cardiomegaly**) can also be present in diseases of the heart valves and congenital heart disease. In addition, enlargement of the outline of the heart can be caused by excess fluid in the pericardium (sac around the heart). If too much fluid is present, it can compress the heart. Usually the most important role of the chest x-ray in the evaluation of heart disease is the condition of the lungs. Fluid build up in the lungs is commonly caused by heart failure. The outline of the vessels near the heart can also be seen on a chest x-ray and can help the physician in the diagnosis of **aortic aneurysm** (localized enlargement in the wall of the aorta) or **aortic dissection** (tearing of the wall of the aorta).

Stress Tests. A stress test is a commonly used procedure to evaluate coronary artery disease. It is also called an exercise stress test or an exercise tolerance test. Stress tests are useful because exercise can reveal abnormalities not seen on an EKG of a heart at rest. A person's EKG, blood pressure, and symptoms are monitored first at rest, then while walking on a treadmill or pedaling a bicycle, and again after exercising. The exercise is gradually increased until a target heart rate is reached (85% of maximum heart rate). If a person is unable to exercise, then medications can be given to stress the heart. If there are deleterious changes of the EKG trace, or effects occur such as chest pain, severe shortness of breath, blood pressure changes or cardiac arrhythmias, then the test may need to be stopped. Even more information can be obtained if certain radioactive materials are injected into a patient during the stress test. These chemicals are then detected by a special camera just after exercise and four hours later. They help to detect areas of decreased blood flow as well as areas where heart attacks have occurred. The effectiveness of the heart's contraction can be determined by measuring the ejection fraction or percentage of blood pumped out of the heart with each contraction.

Cardiac Catheterization. This procedure is the "gold standard" for evaluating cardiac anatomy and function. During a cardiac catheterization, or arteriogram, an X-ray movie is taken to show the anatomy of the coronary arteries, heart chambers, major vessels, valves and congenital heart defects. It is also used to assess the adequacy of heart contractions, pressures in different areas of the heart, output of the heart (ejection fraction), or the amount of oxygen and carbon dioxide in different areas of the heart. After sedating the patient, the cardiologist places a catheter (small tube) into an artery (to look at the left side of the heart) or vein (to look at the right side of the heart) in the upper leg or arm. The catheter is advanced to the heart, dye is injected into the catheter and a movie of the heart in action is taken at different angles to see if any blockages are present in the coronary arteries. A similar procedure known as angioplasty can also be used to treat blocked coronary arteries. A small balloon is inserted into an artery or vein, and then it is blown up at the site of the blockage to create a larger opening. These procedures are invasive and have rare complications such as heart attacks, strokes, abnormal heart rhythms, allergic reactions, perforation (puncture of a blood vessel or chamber of the heart), and even death.

Blood Tests. These are the most important of all the clinical tests when tackling heart disease. As you may have appreciated by now, not only is blood indispensable to life and living on the one hand—on the other, it is probably the most susceptible to disease and infection in conditions of modern living. But the most fascinating part of clinical testing is that the blood “tells all”. There isn’t a single malady that can escape detection when a blood test is called for, regardless of where in the body it may be present or which particular organ is being checked for whatever. From a pathological standpoint, we haven’t even reached halfway in our interpretations of the blood’s “messages” and almost every day something new crops up in our ongoing investigations of the human body—thanks largely to the blood and its constituents which can clearly indicate “when”, “where” and “how much”. You will, therefore, understand the importance of correctly interpreting what the blood is “telling” us when we’re confronted with some of the deadlier diseases.

A heart attack occurs when a coronary artery is blocked, causing a lack of blood and then death to the heart muscle served by that artery. When the muscle dies, it releases certain chemicals present in those muscle cells into the blood where they are retained. These chemicals can be measured and, if present, indicate evidence of a heart attack. The amounts of each chemical, namely CPK, CPK-MB, Troponin and Myoglobin, become measurable and persist in the blood for different periods of time. The following table shows the behavior of these chemicals in the blood.

Lab Test	Begins to Rise	Peak Duration	Found in
CPK	4-8 hrs	48-72 hrs	Heart, Brain, Skeletal Muscle
CPK-MB	3-4 hrs	12-24 & 48 hrs	Heart
Myoglobin	1-2 hrs	4-6 & 24 hrs	Heart, Skeletal Muscle
Troponin	3-6 hrs	12-24 hrs & 1 wk	Heart

There are many other characteristics of blood chemistry, which were discussed in detail in Chapter 2, ISN'T CHEMISTRY FOR THE LABS? Let's review the important aspects of those blood tests that reveal dysfunction and are meaningful from the preventive measures aspect of heart disease.

Cholesterol. This is a waxy, non-soluble substance that is manufactured by the liver and is needed by the body for "cleaning up services". The body manufactures all of the cholesterol that it possibly needs. Excess cholesterol occurs because modern man continues to consume a great deal of saturated fat and high-cholesterol foods. This excess cholesterol combines with calcium and other substances to form the plaques that ultimately block arteries. High total cholesterol levels in the blood double the risk of heart disease.

HDL Cholesterol and LDL Cholesterol. High Density Lipoprotein cholesterol (HDL) is the "good" cholesterol in the blood which keeps the arteries clear by mopping up waste that may have been deposited. These waste products are cycled through the liver and removed from the body. Low Density Lipoprotein (LDL) is the "bad" cholesterol and the body's worst litterbug as it drops all sorts of junk along the arterial wayside. Some folks find it helpful in remembering which is the good and which is the bad by referring to "Happy" cholesterol (HDL) and "Lousy" cholesterol (LDL). Raising low HDL levels and lowering elevated LDL concentrations reduce the risk from coronary heart disease. It is now apparent that other abnormalities of lipid metabolism also play a significant role in the forming of coronary artery disease, and these must be looked for in individuals with otherwise unexplained premature coronary atherosclerosis. Among the patterns associated with increased atherosclerosis are elevated levels of apolipoprotein(a) and of small, dense LDL lipoprotein particles. These lipoproteins and their accompanying lipids appear more likely to pass into the vessel wall and, therefore, to be more difficult to clear. There are other abnormalities in the production of various apolipoproteins such as increased levels of Apo(b) and its affiliated lipoproteins LDL and VLDL. Mutations of many other proteins and enzymes involved in lipid metabolism are clinically important.

Triglycerides (TGs). These are malevolent fatty substances, obtained from foods like butter, that are present in the blood to provide instant energy. Excess amounts are stored in the body as fat for future use. Excess triglycerides are ideal targets for free radical attack, which can turn the fat rancid. Excess TGs assist LDL in its destructive ways of dropping "junk" on the arterial wayside. This "junk" ultimately damages cell linings and forms the plaque that narrows the arteries.

Lipoprotein[a]. Also simply known as Lp(a), this is the nastiest cousin of LDL as it is responsible for thickening the blood and is also ten times deadlier in causing clogged arteries that lead to a heart attack. It is a deadly genetic factor to which Indians are particularly susceptible, as we have the dubious distinction of having the highest incidence rate in the world. Lp(a) is one of the most powerful and most prevalent, independent risk factors for premature coronary artery disease. Elevated levels of Lp(a) start soon after birth or nearly two decades earlier than most other risk factors. Serum levels of Lp(a) have been shown to correlate well with the presence, extent, severity and lesion score on coronary angiography as well as with the occurrence and recurrence of myocardial infarction and cardiac death. High Lp(a) levels in the blood predict coronary atherosclerosis in cardiac transplant recipients, stenosis of the saphenous vein graft following bypass surgery, restenosis following angioplasty and persistently occluded coronary arteries following a myocardial infarction. Premature atherosclerosis of the carotid and femoral arteries as well as aortic atherosclerosis and aortic valve sclerosis are strongly associated with high Lp(a) levels.

Homocysteine. This is a by-product of protein metabolism and to date little is known about this substance, other than that it is a risk factor for life-threatening heart attack and stroke. Normally, it is converted into a harmless substance and quickly cleared from the blood. Unfortunately, there are several factors that can interfere with this process and cause the homocysteine levels to rise to a point where the effects on the body can be lethal. It dramatically raises the chances of a heart attack or a stroke because it hardens and clogs the arteries—including the arteries that feed the heart and brain—and promotes the formation of blood clots in both veins and arteries. Unlike high cholesterol, which may take years for its adverse effects to come to the fore, homocysteine works relatively faster and high levels of this substance are also capable of playing havoc with the circulatory system, besides causing other extremely serious health problems. Homocysteine levels do not fall by themselves but increase as we age.

Taking into account all these serious consequences of heart disease, in our opinion every Indian above the age of 35 years should go through the entire gamut of a ***complete health check-up at least once every year***. Complete health check-ups are easily available and offered by many hospitals and generally cover all blood tests and other non-invasive tests.

Cancer - The Snake In the Grass

This disease is second only to heart disease as the most common cause of death the world over. Though it plays second fiddle to the deadlier, *AND*—from the risk point of view—the more easily contractible heart disease, cancer has successfully held centre stage in terms of fascination, both for the medical profession as well as for the general public. It's quite true that almost every week, and sometimes every day, there's a cancer headline in the news somewhere or other. There probably is no other disease that has received as much scientific attention or research funding as has cancer and it continues to "dominate the limelight". Despite the interest shown in and the amount of research that has been undertaken for cancer, ***the* CAUSE *of most cancers still remains relatively unknown.*** Even though mountains of information are available on this peril, a concrete answer to the problem continues to remain elusively out of sight. It is only recently that ***mutations in DNA sequences*** *and* ***deletion of tumor suppressor genes*** have been linked to abnormal cellular proliferation. Whatever the causes may be (a trip back to the sections on Free Radicals, Chapter 4, and DMSO therapy, Chapter 6, will definitely enlighten you!), the basic reason for a cancer to raise its ugly head is a cell having gone haywire in its reproductive process. The innate quality of a cell is to be able to reproduce other identical cells in order that the life and living processes continue in an orderly and pre-ordained manner. To put the cancer complication as simply as possible, try and imagine your DNA producing genetically "correct" cells somewhat akin to what comes off the assembly line in a modern car manufacturing factory. Somewhere along this production line an extra screw is inserted erroneously and since everything is computerised and automated, each succeeding model at this point will also receive an extra screw. In the case of the human cell, this sort of an "extra screw" will result in chaos as that particular cell will now continue to reproduce with the abnormality amid the profusion of other cells which are reproducing normally. Invariably, the defective cell proves to be the stronger and destroys the normal, healthy cells in the course of its own proliferation. This further leads to dysfunctions of vital organs, disturbances in the circulatory processes and the 'infection' spreading uncontrollably.

Unfortunately, what exactly ***causes*** the cell production line to go haywire is something that is still to be figured out. Knowledge of some of the probable contributing factors is definitely there, but the actual process remains a mystery. The good news regarding cancer is that the instances have shown a slight decline over the past few decades. This could be attributable to changes in lifestyles, improved prevention techniques, early detection and timely treatment, and the availability of more effective medication. However, in a developing nation such as India—where more than 70 percent of the population still remains uneducated and continues to reside in rural areas far from the reach of modern medical facilities—there is certainly the possibility of many incidences of and deaths from cancer remaining undocumented and unreported. Add to this the prohibitive costs of diagnosis, treatment and therapy which even the urbanized section of lower middle classes cannot really afford, and one can comfortably conclude that the overall picture of "cancer under control" in India may not be as rosy as it has been made out to be.

Cancer is a chronic and very complicated disease that requires equally complicated diagnosis, prognosis and treatment. This is because it can manifest itself at anytime, anywhere and at any point in the human body, including the blood and other bodily fluids. Its effects and treatment would differ depending on where exactly and with what severity it has struck. Cancer is not a single disease but is a general term for several distinctly different problems—all of which involve the abnormal proliferation of cells and the destruction of normal cells and tissues. It generally results from genetic and other mutations which may be due to environmental exposure, genetic susceptibility, infectious agents, physical agents such as ionizing radiation and ultraviolet light, hereditary and many other factors. Another point is that ***cancer is* not *contagious*** and is rarely one of those overnight happenings. Many forms of the disease take years to develop, with the affected cells passing through many evolutions before becoming outright malignant and posing a threat. Some of the most common cancers that afflict us humans are **Prostate, Lung, Breast, Colorectal, Uterine, Ovarian, Lymphoma, Melanoma, Kidney, Pancreas, Bladder, Leukemia** (tissues and blood)**, Kidney, Liver, Stomach and Oropharyngeal**. Since the brain, the bones and bone marrow and the superior vena cava of the heart are also "soft targets", there's really no area of the body that can lay claim to being "left untouched"!

There are almost as many alleged risk factors for cancer as there are forms of cancer. Some of these factors could be in doubt, but a majority of them have been so overwhelmingly substantiated that we have no choice but to acknowledge that ALL of us face some degree of risk from unknown quarters. A cancerous tumor has no nerve endings of its own and, therefore, the cancer is very likely to remain unnoticed or unfelt until it begins to adversely affect the immediate nearby tissues. Then there are cancers, like leukemia, which don't involve tumors and may have been around for quite some time before they are eventually detected. Even when symptoms present themselves, each cancer invariably has a different set of symptoms; thus it will be impossible to cover each and every aspect without thoroughly confusing you! If left undiagnosed and untreated, most cancers will result in pain, suffering and ultimately death. To make matters somewhat easier, we will describe some of the bigger risk factors and a few general symptoms which should be reported to your doctor for further investigation.

Risk Factors. All the risk factors enumerated below are very common and are also factors which should be considered for maintaining your overall good health!

- **Smoking.** Smoking cigarettes is akin to pouring boiling oil onto the cells of your lungs—a sure-fire method of ensuring that these cells die prematurely! It has been scientifically proved, beyond any doubt, that all forms of tobacco are linked to approximately one-third of all cancer-related deaths.

- **Diet and Nutrition**. Eating certainly sustains life, but certain eating habits can also kill you! Be careful to avoid stuff like fats, excessive red meats, excess sugar intake— all of which promote cancer. High fibre diets that include plenty of grains, fresh fruits and vegetables are eminently more suitable for positive effects.

- **Alcohol**. Heavy intake of alcohol is a deadly precursor.

- **Obesity and Lack of Exercise**. Obesity is a major problem because fat cells are rich in estrogen. Excess estrogen is a proven factor in breast, ovarian and other common cancers. Obesity can be overcome through a regulated diet and regular exercise. Exercising is much more than just a means of reducing weight and burning off unwanted fat—even those who are not overweight can ward off the risks of cancer through regular exercise. And, the previous two factors, *Diet and Nutrition* and *Alcohol,* often contribute to obesity—a real synergistic combination!

- **Genetic Inheritance.** Hereditary predisposition to some cancers has been linked to events within a gene and is manifested by a strong family history of a relatively rare cancer. Without going into complicated details, suffice it to say that clinical tests have proved that it is possible for cancer-afflicted parents to pass on genetically mutated cells to their children. An American study has revealed that 5-10% of all breast cancers occurring in women under 30 are due to inheritance of an abnormal gene.

- **Environmental and Occupational Exposure**. These are the new risk factors of modern day living. Exposure to high concentrations of certain chemicals and toxins can definitely cause cancer as also can exposure to various kinds of radiation including sunlight.

Symptoms. All the symptoms enumerated below are extremely common ones and may not be related to cancer at all. However, because cancer is so serious and also because early detection means a more effective treatment, diligently exercised personal care will certainly pay rich dividends on the health front. If you experience any of these symptoms, *consult your doctor without delay and get the peril out of the way:*

- Any lump or thickening anywhere in the body, but specifically in the areas of the breast and scrotum.

- A cough that does not go away.
- Any change in a mole or the appearance of irregularly shaped mole-like or scab-like patches on the skin.
- Any unusual bleeding or discharge.
- Unexplained changes in vision, balance or coordination.

DIAGNOSIS OF CANCER. There are so many forms of cancer that the diagnostic procedures involved are just as varied and lengthy. It is not possible to single out a procedure that is common to all varieties of the disease. However, this should not be cause for alarm. Modern diagnostic procedures, for practically every life-threatening disease and very specifically for cancer, are so advanced that physicians are invariably able to isolate the type, location and extent of the disease with almost pinpoint accuracy. However, the diagnosing processes could be cumbersome and lengthy, and sometimes even painful, because cancer is a complicated malady. That is one reason why the emphasis is laid on early detection. Since cancers have a nasty habit of lying hidden and dormant, it is essential that a person understands and accepts that he or she is a potential target. These days free cancer clinics regularly operate all over India and carry out basic tests to detect the presence of some of the more common and malignant forms of this dreaded disease.

Diabetes Mellitus – The 'Sourness' of Sugar

At the outset, it would be entirely appropriate to wish you, dear Indian reader, a hearty "Fare-Thee-Well" as a fellow-resident of the Diabetes Capital of the world! What? You had absolutely no idea that India has the dubious distinction of having the highest incidence of diabetes the world over? Well then, it's about time that you upgraded your knowledge on this aspect—and came to grips with the fact that *YOU* are also probably a prospective candidate! With our totally convoluted eating habits, our over-fondness for 'mithai' (sweets) and everything cooked—nay "soaked"—in pure ghee or vegetable oils and our horrendous approach to hygiene and sanitation, it's really no wonder that we harbour, on the national and collective levels, a pathetic understanding of health and health-related problems. This sort of attitude is abysmally the pits, leading more of us to almost daily fall prey to diabetes which, at the present time, remains an incurable disease albeit not a contagious one.

So what is diabetes anyway? It is a condition of insufficient insulin. The disease develops due to the body's inability to convert the food one eats into the energy required. Food is normally converted into **glucose** (a form of sugar) that is used by the body's cells as a source of energy. Glucose, in turn, increases one's blood glucose level. This increase in the blood glucose level is a signal for the release of a hormone known as **insulin,** which is stored in the **pancreas** within a cluster of cells called the **islet cells**. Insulin's job is to regulate the glucose level in the blood besides assisting in the storage and utilization of glucose within the body. When the amount of insulin released by the islet cells falls below what is actually required, there is not enough 'power' to transfer the glucose from the blood to the appropriate cells. As a result, instead of being used up by the cells, the glucose builds up in the blood. This leads to the situation commonly known to all as "too much sugar". Diabetes is not a disease of blood sugar alone. Unlike most other diseases, it ***involves the entire body and all its systems,*** thus making it a malady that presents serious later-life consequences if not detected early and attended to promptly. It weakens veins, arteries and capillaries besides being a cause for infections of the legs and feet. In severe cases, it can cause gangrene in the extremities. Hemorrhages in tiny capillaries can destroy vision while kidney damage can be serious enough to require dialysis.

Basically, the disease has two names. **Diabetes mellitus** is a condition characterized by an excess of sugar in the blood and/or urine. **Diabetes insipidus** is characterized by an excess of fluid loss by the body. In this section, when we refer to *diabetes* we will only be talking about ***diabetes mellitus***. Diabetes is a complex disease whose chemistry changes from day to day and the daily stress that it puts on the body becomes even greater when there is no control. Not all cases of diabetes are alike and yours may be distinctly different from that of your friend's. Also, there are many individuals currently walking around who are afflicted with diabetes but who have absolutely no idea that they are "ticking time bombs". There are several types of diabetes and within these types the course and severity of the disease varies.

Types of Diabetes. The following types of diabetes are commonly recognised the world over and have been classified as such:

- **Type I (also known as Juvenile Diabetes and Insulin-Dependent Diabetes Mellitus [IDDM]).** This is the most serious form of the disorder and requires lifelong medication and management. It can occur at any age but most commonly starts during youth. In these cases, since little or no insulin is produced by the pancreas, patients are heavily dependent on outside sources of insulin. Prior to 1921, children with Type I diabetes had a very short life expectancy. These days the disease can be totally controlled with insulin, proper diet, exercise and careful monitoring.

- **Type II (also known as Adult-Onset Diabetes and Non-Insulin-Dependent Diabetes Mellitus [NIDDM]).** This is the more common form of the disease which is far more responsive to intervention. This category of the disease is further divided into two subtypes, namely ***obese non-insulin-dependent diabetes*** and ***non-obese non-insulin-dependent types.*** NIDDM is less severe than Type I and starts more slowly. In this form, the pancreas produces some, or at times even excessive, amounts of insulin. However, it is either not enough for proper function or is not being produced quickly enough to effectively influence glucose levels in the blood. This type usually occurs in middle or older age and frequently in overweight people. There are many who have successfully eliminated or greatly improved the condition by losing weight, monitoring diet and making other lifestyle changes. Some instances may require oral medication as well; however, it is possible to control the situation in this form by diet and exercise alone.

- **Impaired Glucose Tolerance.** This is a condition where the fasting plasma glucose level is normal but after glucose intake the levels become abnormally elevated. While people who have impaired glucose tolerance do have elevated blood glucose levels after meals, ***they are not necessarily diabetic*** as their fasting blood sugar is normal. However, people with this condition are prime candidates for eventually developing active diabetes. It is essential that the situation be monitored regularly and the necessary steps initiated because ***at this stage diabetes is preventable.***

- **Gestational Diabetes.** This term is only used to refer to diabetes that develops during pregnancy.

- **Other Types.** There are some other types of diabetes associated with certain conditions and syndromes such as diabetes induced by drugs or chemicals, diabetes secondary to pancreatic disease and diabetes secondary to endocrine disease. These types will not be covered here as they do not occur as frequently as those already mentioned.

TYPE I	TYPE II
Frequent urination	Excess weight
Increased thirst	Drowsiness
Unusual hunger	Blurred Vision
Weight loss	Tingling and numbness in hands and feet
Irritability	Skin infections
Weakness and fatigue	Slowly healing cuts (especially on feet)
Nausea and vomiting	Itching

Diabetic Symptoms. Your first line of defence against diabetes is to *recognise the early symptoms* and have these properly diagnosed. *In the early stages of the* disease, *there are many who either do not notice or fail to recognise any of the symptoms* and live with the disease for some time before overt manifestations start presenting themselves. Many of the older adults may have no symptoms at all other than a vague feeling of not being well. If you experience any of the symptoms listed in the adjacent table, please consult your doctor for in-depth clinical investigations very soon.

Diabetics are extremely prone to problems with the blood vessels and complications in this particular area are frequent. These conditions appear earlier and advance more rapidly than in non-diabetics. Since both large and small blood vessels are affected, complications such as hypertension and atherosclerosis seem to be the principal problems with which to contend. Vascular disease (disease of the blood vessels) is another problem that diabetics may have to contend with as this causes further complications of the circulatory system, including heart attacks. Because of changes in the arteries due to diabetes, some patients experience peripheral circulatory disturbances, especially in the legs. Since blood vessels serving the kidney are often affected, kidney disease and recurrent infections of the urinary tract are common additional ailments in many diabetics. Vascular changes can also affect the eyes where blood vessels are very tiny and fragile, and it is already an established fact that diabetes is the leading cause of blindness. Neuropathy (damage to the neural pathways) is another long-term complication, the most common form being the one that affects the legs causing numbness, tingling and sometimes severe pain and making a diabetic's feet very vulnerable to any kind of injury. The disease is also responsible for worsening the already bad effects of joint, skeletal, muscle and gastrointestinal tract problems.

Another important aspect of diabetes that needs to be viewed seriously is the likelihood of lipoprotein abnormalities occurring. Circulating lipoproteins are just as dependent on insulin as is the plasma glucose. In Type I diabetics, moderately deficient control of hyperglycemia is associated with only a slight rise of LDL cholesterol and serum triglyceride levels with very little or no change at all in HDL cholesterol. Once the hyperglycemia is brought under control, lipoprotein levels are generally normal. However, in obese patients with Type II diabetes the distinct characteristics of the insulin resistance syndrome are the features that involve high triglyceride levels, a low HDL-cholesterol ratio and a qualitative change in LDL particles. Smaller, dense LDL particles are produced whose membranes carry supranormal amounts of free cholesterol. The fact that a low HDL-cholesterol condition normally precedes macrovascular disease should not be overlooked either. Since primary disorders of lipid metabolism may coexist with diabetes, there is a likelihood of lipid abnormalities persisting even after normal weight and normal blood glucose have been restored

What causes diabetes? Medical science has succeeded in determining some factors that largely contribute to the development of this disease but, unfortunately, the exact cause continues to be elusive. One such factor is heredity. Diabetes can be passed on through genes, but a child of non-diabetics can also become a diabetic since the disease may skip generations because of a genetic coding which prevents it from appearing in every generation. Stress which affects the cells of the body appears to be another reason why diabetes strikes certain people. Stress may be physical or emotional, such as surgery or a serious infection or an accident or emotional shock. One such stress is extra weight and it is already a medical fact that obesity, which affects insulin utilization, is a major contributor to the onset of diabetes. In addition, there are many medications that affect the body in stressful ways as well. Pregnancy is another stress that the body has to endure and many women in this condition contract diabetes.

DIAGNOSTIC ASPECTS. As with all other diseases, a series of clinical tests are required to confirm the presence and severity of a suspected disease as well as to aid in the making of an accurate diagnosis. From the patients' point of view, these tests are fairly simple and involve investigating only the blood and urine to confirm the presence of diabetes. We shall not go over the additional tests that may be required when other complications run concurrent with diabetes, since the investigations carried out for heart and kidney diseases normally cover the entire spectrum. Basically what you need to know is whether high levels or accumulations of glucose persist in either the blood or the urine. This will indicate that the glucose is not being channelled properly and that the insulin being produced by the body is either insufficient or is not being used in the correct manner. Because insulin is not performing its job of removing glucose from the bloodstream and transferring it into the cells, a time comes when the blood glucose becomes so high that the kidneys start to overflow this glucose into the urine. In addition to measuring the glucose level, the tests will also indicate the presence of ***ketones***. Without insulin, fat stores in the body break down and substances known as ***ketones, ketone bodies*** or ***acetone*** are produced. Being acidic in nature, these create problems when stored and accumulated in the blood. At high enough levels, ketones are also capable of overflowing into the urine.

Blood Tests. Blood glucose tests will indicate whether or not there is either too much sugar or not enough sugar in the blood and whether ketones are present. The concentration of blood glucose is an accurate yardstick for indicating the presence of diabetes as well as for exercising diabetic control. The specific investigations are carried out as follows:

1. **Blood Sugar**. For this test, first a fasting (normally no food after the previous night's dinner) sample of blood will be drawn. Evaluating this sample will indicate the extent to which carbohydrates from your last meal have been removed from the bloodstream. Immediately after this sample is drawn, you will be asked to consume a normal meal. Two hours after this meal a second sample of blood will be drawn. This sample will show the blood glucose concentration and is an indication of the glucose absorption efficiency of the body.

2. **Glucose Tolerance Test**. This test is somewhat similar to the blood sugar test, in that first a fasting sample of venous blood is taken. Following this you will gulp down 75 gm of pure glucose diluted in 300 ml of water. Three more venous blood samples will be drawn after one-hour intervals. The measurements of each sample will be ascertained and plotted on a graph which will display an accurate portrayal of just how well your body handles glucose. This test is very valuable because it can confirm the presence of ***impaired glucose tolerance***. Persons

suffering from this condition are more susceptible to developing full-blown diabetes than some others.

3. **Glycosylated Hemoglobin.** This is another test which clearly indicates the concentration of glucose in the blood. Glycosylated hemoglobin is created when glucose is attached to hemoglobin cells (red blood cells) and the concentration of these molecules, being higher in diabetics as well as those whose diabetes is poorly controlled, is a good barometer of average glucose content. Depending on its concentration in the blood, glucose attaches itself to the hemoglobin slowly. As the life span of a red blood cell is about four months, a high concentration of glycosylated hemoglobin means that the condition has been building up over a period of time. Because this test is indicative of what occurred previously rather than what is taking place now, it is useful as a long-term monitor as well as helping to determine the correct course of treatment.

The following table will help you to understand what some of these diabetic blood tests are all about and what the readings mean.

Condition	Normal	Normal Glucose Tolerance	Impaired Glucose Tolerance	Diabetes Mellitus (II)
Fasting plasma	065-100	<110	110-125	≥126
2 Hours after meals or glucose load	100-120	<140	≥140& <200	>200

NOTES:
Figures in mg per ml of blood
< indicates "less than"
> indicates "more than"
≥ indicates "equal to and greater than"

If you ***do not*** have diabetes or impaired glucose tolerance, the resulting curve when plotted on a graph will show a normal pattern. If you ***do*** have *impaired glucose tolerance*, the graph will be similar to that of a non-diabetic's in pattern but will indicate higher blood sugar concentrations with normal fasting or end-of-test levels. If you ***have*** diabetes, the graph will indicate an ever-rising blood glucose level that shows no indication of dropping even by the end of the test.

Urine Tests. Urine testing is normally part of a complete, routine physical examination. Being basic diagnostic tools, urine tests provide valuable information on just how well the body is functioning. Though these tests do give positive indications when testing for sugar content, the readings may not always accurately reflect the level of glucose in the blood and they are harder to interpret than blood tests. Two types of urine tests, *semi-quantitative* and *quantitative*, are resorted to when investigating diabetes. In the semi-quantitative type, only a small sample of urine is given and the test is easily and speedily accomplished. The test involves mixing the urine sample with certain chemicals so that a colour change occurs. The hues will vary in tone as per the amount of sugar and ketones concentrated in the sample. A patient can perform this simple test at home. In the quantitative measurement type, though a slightly cumbersome procedure, the advantage is that only glucose and ketones are measured and that too in an exact concentration rather than in a range. In this procedure the urine may have to be collected at timed intervals over a 24 hour period. As indicated, the urine sugar test is only a rough guide. The blood always contains some amount of sugar (between 80 to 140 mg/ml) round the clock, since certain tissues, like those of the brain, cannot survive without readily available glucose. When the blood sugar exceeds about 180 mg/ml, the sugar will appear in the urine. If no sugar is indicated in the urine, it is a rough indication that at the time of taking the urine sample the blood sugar was less than 180 mg/ml. Blood sugar is definitely the more accurate parameter for assessing control of diabetes.

Though incurable at this moment in time, diabetes—if correct and conscientious control and treatment regimens are exercised religiously—as a disease is neither fatal nor debilitating. However, since it is such a complex disease, even the smallest of lapses or carelessness could prove otherwise. Diabetes is not a disease of the blood sugar alone. It involves your entire body and is associated with long term complications involving almost all vital organs such as the brain, the heart, the kidneys, the eyes and the nervous system. These complications will definitely occur earlier in those cases where there is a glaring lack of adequate control over the disease. In the Indian context, it is quite true that for every known diabetic, there is a diabetic waiting to be discovered! Since the body is also a complex machine, it will work at optimum levels of efficiency only when we have a basic understanding of all its parts and coordinate the various functions. The individuals who live the best, have the fewest complications and exercise the best control over this disease are those diabetics who have centred their active lifestyl health habits and dietary control as well as proper medication. Controlling diabetes on a day-to-day tremendous self-discipline, but that's an eminently better scena… having a disease that is untreatable. It is also a very personal disease and an individual must garner the best possible understanding of how to live with it. The 'good thing' about diabetes is that it is likely to ***prolong life!*** Patients have to improve their basic diet and care, so they automatically reduce many of the problems they might have had without the self-discipline imposed by diabetes. Many have actively reduced excess weight, stopped smoking, started exercising regularly and stuck to the correct foodstuffs thereby living far improved and healthier lifestyles—*benefits that probably many would never even have tried to achieve if diabetes had not come knocking at their door.* It should be remembered that, though a physician's advice and guidance is essential when tackling diabetes, the most important factor in controlling this disease is the individual patient himself.

Renal (Kidney) Disease – 'Wasteful' Expenditure

The kidneys, like all the other vital organs, require nurturing and care. Unfortunately, these are generally considered minor organs—unwittingly, of course—whose sole function is the production and excretion of urine. In addition, even their exact location is unknown to many! Everyone is born with two working kidneys. These are bean-shaped, about the size of a human fist, and are located on either side of the spine, on the rear of the body, at about waist level. An amazing aspect is that a person requires just *one* working kidney to remain healthy —in fact, if only 70 percent of that ***one*** kidney is functional, good health can still be maintained!

The main task of the kidneys is to purify the blood. As you are well aware by now, blood circulating through the body carries substances and nutrients that are essential for life besides also carting along a lot of waste matter and excess water that needs to be ejected. Firstly, the kidneys ensure good health by separating the waste matter—which is then excreted out in the urine—from the essential substances, thereby keeping the blood chemistry 'in balance'. These organs are comprised of millions of minute filters. The cleaning process takes place when arterial blood enters the kidneys and passes through these tiny filtering units called nephrons (see Figure 11). These resemble tiny funnels with long stems, each having two twisted tubes. . Matter that is not needed by the body is separated at this point and only healthy, nutritious blood returns to the bloodstream. The waste products and extra water mix to form urine, which passes from the kidneys, through tiny tubes, into the bladder. ***Urea*** and ***creatinine*** are the most common forms of waste removed in the urine. These two substances are normal constituents of blood, though in small quantities (refer to Chapter 2). If the quantities are too high, there is every likelihood of inadequate kidney function. Measuring the amounts of urea and creatinine in the blood gives a clear indication of the functioning efficiency of these organs.

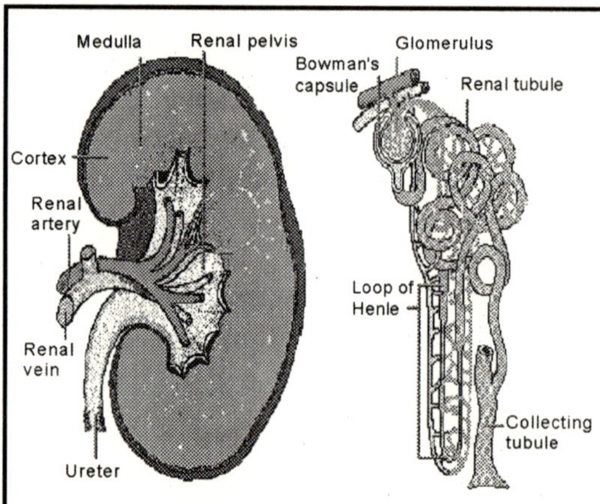

Figure 11 – Details of the Kidney (left) and Nephron (right)

Probably the most important job of the kidneys is the production of a chemical, or hormone, which aids the body in the creation and production of red blood cells. A low red blood cell count results in the body experiencing abnormal fatigue, so the kidneys are relied upon for the production of the right amount of red blood cells to keep the body strong and healthy. The kidneys are also the “fluid controllers” of the body. They can sense when there is either too much or too little fluid in the body and produce the exact volume of urine so that body fluids remain in a balanced state at all times. If one drinks more water than is normally required, the kidneys get rid of the excess. If one tends to sweat excessively, the kidneys ensure that the right amount of body fluids is retained. Besides controlling the amount of phosphate in the blood (too much phosphate will weaken the bones by taking calcium out of them), the kidneys are also responsible for the release of Vitamin D that is used to keep the bones healthy. Another important action of the kidneys is that they keep a check on and control blood pressure. From the foregoing it is quite evident that the importance of the kidneys cannot be understated.

When the kidneys fail to perform all of the above tasks efficiently, the overall health of the body is definitely jeopardized. Kidney disease is a *silent* disease, therefore what needs to be recognised is the various symptoms that present themselves when the kidneys start to fail. One important fact regarding renal diseases needs to be emphasized here—***if contracted, most of the disease conditions are irreversible.*** If caught in time, the various diseases to which the kidneys are most prone can be successfully treated. However, it must be clearly understood that by "successfully" it is meant that medicine can only ***slow down the* progress *of the malady.*** *The disease cannot be eradicated or arrested.* So the earlier the disease is discovered and treatment initiated, the reaching of an end-stage renal disease condition will be put off by that much longer! Some of the common symptoms that present themselves when disease is imminent are:

- ***Feeling unwell or sick*** Because extra waste matter is not being efficiently and speedily removed from the system, this build up causes one to feel sick. Extra urea in the blood brings on uremia, which results in drowsiness, a confused mind and nausea.
- ***Paleness/tiredness***: Since the kidneys cannot instigate the required red blood cell production, the body is not getting enough energy which, in turn, brings on extreme tiredness and facial paleness.
- ***Skin puffiness:*** As the kidneys are unable to remove the extra water, fluids start collecting in the bodily tissues. This results in skin swelling in the ankles or around the eyes. This complication is known as uremia. If fluids happen to seep into the lungs, one would feel breathless as well.
- ***Passing urine:*** Although kidney malfunction may have already set in, the body will continue to produce urine. This urine is not normal however, because it does not contain all

the waste matter that should be a usual constituent; thus the amount of urine being passed is not a measure of the healthiness of the organs.

- ***Foul breath & taste***: As the kidneys are unable to extract the extra urea that may have built up, this will cause bad breath and a loss of appetite.

There are a number of reasons why the kidneys stop functioning properly—***diabetes, glomerulonephritis, high blood pressure*** and ***polycystic kidney disease*** being some of the more common causes. Renal disease is established in two ways: discovered incidentally during a routine medical evaluation or when specific evidence of renal dysfunction—like hypertension, edema, nausea and hematuria (the presence of blood in the urine)—is investigated clinically. Renal disease may be ***acute*** or ***chronic***. ***Acute*** renal failure is a worsening of renal function over hours to days resulting in retention of nitrogenous wastes and creatinine in the blood. ***Chronic*** renal failure results from a loss of renal function over months to years. We shall now focus on some of the specific renal diseases and how they present themselves.

Acute Renal Failure. This is defined as a sudden decrease in renal function which results in an inability to maintain fluid and electrolyte balance or to excrete nitrogenous wastes. A sudden increase in BUN and serum creatinine will be noted and the patient will generally display nausea, vomiting, malaise and altered sensorium. Hypertension is rare, but the fluid balance will definitely be altered.

Acute Tubular Necrosis. When acute renal failure occurs due to tubular damage, the condition is known as acute tubular necrosis. Two major causes of this condition are ischemia and toxic exposure. The symptoms are almost the same as the above. Renal blood flow is inadequate to maintain the essential cellular formation and substance, which occurs in settings of prolonged hypotension or hypoxemia such as dehydration, shock and sepsis.

Interstitial Nephritis. An inflammation of the kidney crevices, this disease accounts for about 10 – 15 percent of inherent renal failure cases. Besides the kidney tissues becoming swollen with serous fluid, possible tubular damage can also occur. The condition may be brought on by infectious diseases or immunologic disorders or it could occur by itself without any outside cause. Patients usually experience fever, rash eruptions, acute renal insufficiency, pus in the urine, white blood cell casts and hematuria. *T lymphocytes* can either cause poisoning of the cells (cytotoxicity) or the release of *lymphokines* that recruit *monocytes* and inflammatory cells. The good news is that this condition has a good prognosis and full recovery can be achieved over a period of time without the malady progressing to end-stage renal disease. However, the patients will require acute dialytic therapy as part of the treatment.

Glomerulonephritis. This is a rare disease and accounts for only about 5 percent of acute renal failure cases. A ***glomerulus*** is a cluster of small organisms, tissue or blood vessels in the capillaries of the kidney. A person afflicted with this malady will evidence inflamed glomeruli that have suffered some injury. The larger the number of glomeruli involved and the more severe the lesion, the more unlikely it is that the patient will make any sort of recovery. The symptoms of this malaise are hypertension, edema, abnormal urinary sedimentation, dysmorphic red cells, red cell casts, hematuria, mild proteinuria and acute renal insufficiency. There are many other conditions of this disease that come under the heading of ***glomerulonephropathies***, but these will not be covered here.

Chronic Renal Disease. Over 50 percent of the cases of chronic renal failure are due to diabetes mellitus and hypertension. Glomerulonephritis, cystic diseases and other urological diseases account for 20-25 percent. About one-sixth of cases are due to unknown causes. Chronic renal disease is rarely reversible and leads to a progressive decline in renal function. ***This happens even when the main cause for the onset has been eliminated.*** The symptoms of this disease develop very slowly, are generally non-specific and an individual can remain free of any debilitating signs until renal failure is far advanced. Manifestations include ***weakness, fatigue*** and ***bodily discomfort*** or ***feelings of unease*** even without the development of any specific disease. Other symptoms are gastrointestinal complaints such as ***anorexia, nausea, vomiting, a metallic taste in the mouth*** and ***hiccups***. Common neurologic problems include ***irritability, difficulty in concentrating, insomnia, restless legs*** and ***twitching***. ***Pruritus*** (violent itching of the skin) is common and difficult to treat. As uremia increases and advances, ***decreased libido, menstrual irregularities***, ***chest pain from pericarditis*** and ***paresthesias*** can develop. The patient's skin can be ***easily bruised*** and will have a ***yellowish tone***. ***Uremic fetor*** is the characteristic fishy odour of the breath. Evidence of previously elevated BUN and creatinine, abnormal prior urinalysis and stable but abnormal serum creatinine on successive days are most consistent with the chronic process. The complications that arise from this condition are far too many to discuss without totally confusing you and we'll leave those for the doctor to explain if the need arises. Suffice it to say that this malady can have serious debilitating consequences on one's everyday living routines.

Renal Artery Stenosis. The two most common forms of renal artery stenosis (narrowing of kidney arteries) are ***atherosclerotic ischemic renal disease*** and ***fibromuscular dysplasia*** (abnormal growth of muscle fibres in the kidney tissues). Atherosclerotic ischemic renal disease commonly occurs in those over 45 years of age who have a history of atherosclerotic disease. This accounts for between 67-95 percent of renal artery stenosis cases. The risk factors of this malady are diabetes mellitus, tobacco use, hypertension and renal insufficiency. Fibromuscular dysplasia primarily affects young women. Any unexplained hypertension occurring in a young woman should be a signal for her to be screened for this disorder.

What we have covered here are the generalities that come under acute and chronic renal diseases, however, there also exist various sub-heads that are treatment-specific in their own right. Some of these are listed below for your information:

- **Glomerulonephropathies**
  Diseases Demonstrating Nephritic and Nephrotic Components
  Glomerular Diseases with Nephritic Presentations
  Nephretic Syndrome
  Nephrotic Disease from Systemic Disorders
  Nephrotic Disease in Primary Renal Disorders
- **Chronic Tubulointerstitial Diseases**
- **Cystic Diseases of the Kidney**
  Simple or Solitary Cysts
  Autosomal Dominant Polycystic Kidney Disease
- **Medullary Sponge Kidney**
  Juvenile Nephronophthisis-Medullary Cystic Disease

- **Multisystem Diseases with Variable Kidney Involvement**
  Gout and The Kidney
  Multiple Myeloma
  Sickle Cell Disease
  The Kidney and Ageing
  Tuberculosis

It is hoped that the preceding paragraphs have jolted you out of your lethargy and kindled a new respect for your kidneys and that from henceforth you will adopt a different and positive approach to the care of these extremely delicate and vital organs. If "unclean" blood continues to circulate in the body, due to dysfunctional kidneys, a host of other complications are being unnecessarily invited which would compound health problems no end. Again, kidney diseases can precede and lead to urologic problems which are a different kettle of debilitating fish altogether! A recent issue of *The American Journal of Kidney Diseases* reported that it is estimated that one in 9 Americans has chronic kidney disease, but many do not know it! The journal also reported that 300,000 Americans are on dialysis and 80,000 have received kidney transplants. It is hoped your new found respect for your kidneys will keep YOU from being included in these statistics.

It is strange but true that having any heart or lung diseases does not necessarily mean that one will develop a kidney disease as well. However, if one develops a kidney disease, ***it necessarily means*** that this is the precursor to heart disease and many other bodily dysfunctions.

After Word

There are a host of other diseases that can cripple, maim or even prove fatal to human beings. What has been highlighted in the foregoing paragraphs are the most common of the afflictions that come under the heading of "deadly diseases" and the ones that we can contract very easily. Though none of them can be termed infectious even in the wildest sense of the word, the rate at which people are stricken by any one, or various combinations of the four, can lead to the belief that they are indeed contagious.

CHAPTER 11

ALTERNATIVE MEDICINES: TRADITIONAL NATURE CURES

Much Ado about Nothing?

Alternative Medicine may be literally defined as any form of medicine that does not fall under the category of universally acknowledged orthodox medicine, or allopathy, which originated in the West and is currently the primary accepted norm of medical practice. The term '*alternative medicine*' may also be loosely applied as a general description of all forms of medicine that are not labeled under allopathy. At the 1973 convention of the very first World Congress of Alternative Medicines in Rome, over 135 different therapies came up for discussion and more than 100 forms of practices were identified and enlisted by the World Health Organisation (W.H.O.). These were categorized under the general headings summarized in the following table.

Traditional Medicine	Those being practiced from time immemorial.
Complementary Medicine	Those that supplement orthodox medicine in many cases.
Unorthodox Medicine	Those that specifically differ from orthodox medicine.
Holistic Medicine	Those that treat the human body and its component parts and senses as a single entity.
Ethno Medicine	Those traditional practices which are closely associated with the norms of living and culture.
Natural Medicine	Those that derive their roots from the laws of nature and natural elements/substances

The alternative system of medicine had a tremendous beneficial impact on maintaining health in ancient civilizations. Unfortunately, the proven therapeutic efficiency and values of the system have been swallowed up in the mists of time. These were everyday practices of ancient days and a scientifically viable approach was applied to every possible malady to ensure positive results. It would currently require an entire process of rediscovery and in-depth study if the present and the future are to reap any real benefits from the alternative systems. India has been a world leader in the propagation and development of various systems of medicine from the time when allopathy was either not even thought of, or was just in its embryonic stage. There is an urgent need to preserve and propagate these systems once more by using the advanced techniques of modernization in the fields of documentation, research and teaching. Controlled scientific research, with regard to the efficacy and side effects of alternative medicine, is absolutely essential if the system is to secure a firm base and gain the popularity and trust that are essential. We must judiciously combine traditional knowledge with modern usage to come up with a universally acceptable and foolproof remedy for the various, till now, seemingly incurable ills that currently plague the modern world.

Propagating the Practice; What's Up Doc?

Although some of the Indian systems are considered extremely "old hat", no one can deny that, despite its modernity, undue importance and fashionable façade, allopathy is fast losing out in terms of reliability. If our practices are to be uniformly and globally standardized and accepted, we need to initiate immediate efforts to recognise, promote and bring about a resurgence of traditional medicines at the national level. Governments must include alternative medicine in national health policies and encourage traditional healers to come forward for centrally sponsored orientation programs. These would provide training in modern methods and education on how the integration of traditional and orthodox systems could be successfully achieved. Due recognition, at the national level, must be accorded to the alternative systems and funds must be specifically allocated for broadening the spectrum. In addition, it is imperative that the overwhelming importance to and dependence on allopathy be drastically reduced. It is rather inane that one does not wholeheartedly support medical practices that have not only withstood the test of time, but are also proving to be far more reliable and dependable and, ***as a matter of documented fact***, much safer.

Indigenous systems, developed and effectively practiced over the millenniums, still have a lot to offer—***an advantage that must be expeditiously exploited to the maximum limits***. It will be necessary, and also highly desirable, to successfully integrate the scientifically proven components of ancient systems into modern medicine. This can only be achieved if the revival studies and thorough research are given the requisite fillip. Only then could we be reliably certain that the anticipated cures would actually take place. Doctors **must** be in a position to

guarantee a cure and patients **must** be made aware, not only of exactly what their ailments are but also about the specific course of treatment and anticipated results that the prescribed therapies involve. Today's physicians need to change both their attitude and approach toward alternative medicine. Being dogmatic and adopting rigid anti-stances, ***for purely selfish and financial reasons***, is not doing the orthodox medical fraternity any good. There is so much more that orthodox medicine is truly in a position to offer, but only if it openly and supportively allows itself to be complemented and supplemented by the alternative systems. There are many diseases such as heart maladies, peptic ulcers, bronchial asthma, diabetes, kidney disease, and viral hepatitis—to name but a few—that allopathy has confronted with miserable failure ***but where alternative medicine has made an effective and definite impact***. There is no doubt that all of us are desperately and eagerly looking forward to that ultimate and final frontier of "perfect health". This reality can be achieved, but only if all spheres of medicine work in a spirit of unison and cooperation with just one aim in sight. Governments must support and encourage such an aim by, firstly, putting all the preconceived notions and misconceptions about alternative therapies on the back burner and, secondly, by pursuing the ultimate objective with foresight, vigor and dedication.

What's the Hole in My Pocket Got to Do With It?

In recent times there has been a marked increase, both in India and abroad, of rekindled interest in alternative medicine. After years and years of completely relying on and pursuing the dynamic science of allopathy, an increasing number of patients the world over are turning back to***, and depending a lot more on***, traditional practices to relieve or cure their many ailments. The prime-mover of this international reversal is the growing realization, not to mention the resulting scare, of the ever-increasing adverse side effects of allopathic medicine. Secondly, especially in the context of the developing countries, the soaring costs have literally put allopathy beyond the reach of the common man. On the third front is the frustration that has been generated by orthodox medicine's abject failure to affect the desired cures. Now that the interest in traditional medicine has staged a comeback, it may be easier to further the prospects, ***but only if all the man-made and artificial misinterpretations are also done away with***. Though the numbers are small and practically inconsequential to the overall effects, there **do exist** a few orthodox practitioners who have the courage to tell a patient when western medicine can no longer help, and who even go to the extent of suggesting that alternate therapies be approached as a last ditch measure. Then there are those orthodox physicians whose routine practice includes a combination of the western and traditional systems and who also lend their unqualified and vociferous support to alternative therapies. These doctors are possibly the only true 'healers' since they go to every extreme to exploit the maximum therapeutic benefits from all systems of medicine. We could certainly do with a lot more of this particular ilk and if only their numbers were somehow to increase, we would definitely witness a new dawn much sooner rather than later!

The renewed interest in alternative medicine must be kept alive and thriving. It is a fairly easy task, especially in the developing nations, because in many societies the rural folk have cultural links to traditional practices and accept them more readily. Then again, the simplicity of these therapies with assured and positive results and the ***low costs*** all add up to act as a magnet. Also there are those who have a religious bias against and philosophical aversions to western medical practices. All this put together may still fail to achieve the desired goal, unless the orthodox medical fraternity changes its attitudes. Since medicine has only one goal—the ultimate benefit of mankind and other living creatures—it is rather silly and immature on the part of any one system to run down others, ***especially in a situation where all systems are pursuing the same goal!*** The fact that unstinted cooperation can only bode well must be realized and accepted immediately, especially by the proponents of orthodox medicine. Self-serving interests will never, in any manner whatsoever, benefit the general public. International governments must work in concert and heartily propagate alternative practices if the hopes and aspirations, on the medical front, of the global community are to reach fruition.

Are Alternative Medicines Really That Good?

The most important reason for the "superiority" of alternative medicine is that it tackles any and every malady by first considering the human body as a single entity—with all the component parts and senses working in perfect concert as a supplement to the whole. Health may be defined as a state of physical well-being, high mental acumen, and becoming socially adjusted and spiritually developed. Alternative medicine is guided by this definition and gives due importance to both the preventive and promotive aspects of health. Since it acknowledges the existence of many planes, it is invariably able to get to the root of the problem with a higher degree of accuracy. The alternative therapist never groups patients together—***even if they are all suffering the same illness***—because the aspects of cause and effect, or the underlying source, may and will be different for each individual. Therefore there has to be a definitive approach in the eventual treatment—*all primarily to benefit* ***only the patient***. Another reason for this superiority is that the therapies employed are very simple and every system has its own unique philosophy. Potent or hard drugs find no place in these treatments, and the use of any invasive diagnostic or surgical procedures are also totally eschewed. The non-use of drugs automatically ensures that any adverse side effects are either completely absent or of much less severity.

Alternative forms of medicine are the tried and tested medical disciplines of ancient times that were handed down over the eons for our benefit. These treasures have been lost somewhere along the way partly due to Man's stupidity and carelessness and partly because of the human obsession with modernization. We are very fortunate that many such systems, especially in the developing nations, have somehow been able to survive the ravages of time. The need now is to broaden the spectrum and scope of the knowledge provided by these 'survivors.' The most overt of the superior aspects of alternative medicine is ***the ability to cure***. In contrast, all that orthodox medicine has so far offered is ***relief*** from chronic problems—***never*** any cure! In addition to the fact that orthodox medicine is far too expensive in most chronic cases, it is also subject to non-availability and short supply in many parts of the world. ***This is not the case with traditional medicine***. Another important aspect of alternative medicine is that it can and has been used to supplement or complement orthodox therapies with a great degree of success, efficiency and reliability. ***Vice versa is not possible!***

So What Role Do Herbs Play in Modern Medicine?

Though there are a myriad of traditional cures prevalent in the world, herbs play an increasingly significant role in today's scenarios when the damaging effects of food processing and over-medication are literally running rampant. The practice of herbal medicine dates back to the earliest times of recorded human history. The ancient civilizations of India, Greece, China, Rome, Egypt and the Incas have all provided evidence of the profuse use of herbal treatments for revitalization of the body and eradication of diseases. Plants were the chief source of medicine in those times and herbs were credited with mystical and almost supernatural powers of healing. In India, herbs are fleetingly mentioned in the ***Rigveda***, written about 3500 to 1800 BC, and with greater detail in the ***Arthavaveda***. The two most important works of the Ayurvedic system of medicine, ***Charaka Samhita*** and ***Shusruta Samhita***, have recorded comprehensive information regarding herbs and herbal treatments. An important Chinese work, Pen Ts'ao Ching dated 2800 BC, has a compilation of 1000 herbs and instructions on how to use them. Even the Greek father of modern medicine, Hippocrates (460 BC), listed hundreds of herbal remedies.

Herbs are more often used in cosmetics, food and tea than as alternative medicines, but the growing interest in herbs as medicine is part of the fast-changing aspect of lifestyles. The movement is based on the belief that plants have a greater potential in their role as curative substances. Another important aspect is that herbs are credited with a kind of 'everlasting' history and available records definitely indicate that they were extensively used ***to cure practically every known illness***. So, in comparison with the limited efficiency of orthodox medicine, there are absolutely no doubts regarding the efficacy, usefulness and far-reaching beneficial aspects of herbs as curative agents. As with all natural medicines, herbs act holistically on the human body. Since these medications always single out the root of the problem, the healing aspect normally tends to follow a laid down path that methodically affects repairing, fixing and strengthening as it trundles along. The ultimate goal is the restoration of the vital energy that controls all the body's essential functions. This being the case, the importance of herbs in complimenting, and as supplements of, orthodox medicine cannot be understated. In simple terms, complete emphasis centres on the word ***'cure'***.

Modern medicine may and can offer relief, however, it is herbal medicine that will speed the curative processes and bring about the ultimate remedy. Herbs may be used in many different ways, but the prime objective of their utility is the direct interaction with the body's chemistry. Regardless of the manner, forms and purposes of the many usages—food, medicine, cosmetics, fragrances—unless the active constituents of herbs are properly absorbed by the body, none of the required benefits would be derived or achieved. Once the constituents are absorbed into the bloodstream, they will circulate to influence the whole system. The skill of the herbalist lies in maximizing this effect to balance and strengthen the body's own healing mechanism—instead of suppressing or disturbing it, as is the wont and tendencies of most modern day drugs. These active constituents of herbs can enter the body in several ways. Oral consumption will provide a path directly through the digestive system; skin applications through cosmetics or medicinal poultices will be absorbed through the pores; applications on the eyes through lotions and compresses, and through aroma therapy for essential oil absorption. While most herbs have little or no harmful side effects, some may cause some undesirable reactions in particular individuals. Therefore, herbs should be initially tried one at a time, in small doses, and any side effects noted. ***Not all herbal applications are effective in every case in every person*** and these should ***NEVER*** be used as substitutes for professional medical attention in emergencies or serious chronic cases.

Extract and Preserve - Don't Go Overboard!

Many herbs contain powerful ingredients that, if used correctly, can help to completely heal. The early pharmaceutical industry was based upon its ability to isolate these ingredients and make them available in purer forms. Herbalists, however, contend that nature provided other substances in the same herb to balance out the more powerful ingredients. These other components, though they may be relatively less potent, help to act as a buffer, a synergistic agent or a counterbalance when working in harmony with the more powerful ingredients. Therefore, by using herbs in their complete form, the body's healing process utilizes a balance of the ingredients ***as provided and dictated by Mother Nature***. All roots and bark are fungicidal and bactericidal; otherwise pathogens would destroy them in the ground. Roots, bark and other herbs will retain their medicinal value for years when preserved in a thoroughly dried state. As a general guideline, most of the bitter-tasting herbs are medicinal in nature. Those with a pleasant taste are potentially less potent and can be used more often. Certain herbs should only be used during a process of healing and also not for periods longer than six months. Most herbal ingredients are much more potent when taken from freshly collected herbs. The fresh leaves and roots of herbs can be used in their natural form or they may be used in the form of tablets, capsules, liquid beverages, bark pieces, powder, extracts, tinctures, creams, lotions, salves and oils. The whole leaves of many herbs can also be dried and made available to consumers.

Herbs do perform many healing functions in the body but ***they must be used appropriately***, **not indiscriminately**. It must be remembered that

not all plant life is beneficial. There are poisonous plants and some that are deadly—especially if used for long periods of time. In fact, an important aspect is that herbalists use herbs with great care. Since herbs contain active ingredients, one needs to be made aware of the fact that ***some of these elements may interact negatively with prescribed medications***. It would be a wise to consult health professionals when there are any questions about safety. Many people believe that the healing properties of herbs are just as effective as drugs because herbs can be very potent; therefore, it is important to regulate dosages and exercise care as most over-the-counter drugs are too strong. It is a point of interest that, currently, in many industrialized countries, herbs are being prescribed by physicians and prepared and sold through neighborhood drugstores. It is also an established fact that in countries where access to hospitals and doctors is limited, herbal remedies are the only form of medicine readily available.

In my quest for promoting health and a better quality of life, I have attended various medical meetings and seminars on herbs and herbal remedies in the United States and Europe. It was amazing to discover that many American and European orthodox physicians were not only very knowledgeable about the herbs and herbal practices of ancient India, but also actively and successfully integrated these traditional treatments with their normal, everyday orthodox therapies. On returning to India, I successfully incorporated many herbal remedies into the various alternative therapies that are conducted daily at our clinic. Needless to say, this innovation has greatly benefited many of our patients. While on the subject of foreign medical environments, it is worthwhile to make mention here about **Commission E**. This is a famous German government panel that makes an in-depth evaluation of herbal remedies and judges their usefulness. Instead of setting up or running studies itself, this panel gathers and reads all available data on herbs and herbal treatments from all over the globe. Based on the collective findings, the panel then either approves or rejects various remedies. Even though it has no official power outside of Germany, the findings and recommendations of Commission E are extremely useful to the rest of the world as this panel is one of the few reputable review boards that systematically studies and evaluates information on herbal remedies.

Complete information on all the valuable herbs available, and how they can cure some of the deadlier diseases that plague mankind, is far too vast to be covered here. However, I am in the process of compiling a second compendium, along the lines of this book, which will be devoted exclusively to herbs and will serve as an informative guide on the subject.

Other "Happy Faces" of Alternative Medicine

Now we will introduce thirteen of the most common alternative medical practices from among the 100-plus alternative practices listed by the World Health Organisation.

HOMEOPATHY. The homeopathic system of medicine is one of the finest mediums of cure available to mankind. "The highest ideal of cure is rapid, gentle and permanent restoration of the health, or removal and annihilation of the disease in its whole extent, in the shortest, most reliable, and most harmless way, on easily comprehensible principles", so said the originator of this system, Dr. Samuel Hahnemann. Homeopathy achieves this dictum in totality. The system seeks to cure in accordance with the natural laws of healing and uses medicines made from natural substances: animal, vegetable and mineral. Not a single medicine is used without first fully knowing what effects it would have on healthy human beings. The harmlessness of this system cannot be overemphasized in present-day scenarios where indiscriminate usage of antibiotics and other "wonder drugs" on a regular basis cause more problems than cures. The homeopathic approach is that the body is always striving to keep itself healthy and in balance, and the force which acts in this protective manner is the vital force.

When the body is threatened by harmful external forces, the vital force—or defence mechanism—produces symptoms such as pain, fever, mucous or a cough. Though discomforting to the patient, homeopathy views these symptoms as a healthy reaction of the body to harmful forces and seeks to support these rather than interfere with them. Pain is an alert that something is amiss; fever inactivates many viruses that attack the body; mucous carries off irritating material in the respiratory tract and a cough expels this mucous which would otherwise hinder breathing. Orthodox therapies take an opposite view—that any such manifestations must either be opposed or suppressed. Since there are constant and ongoing changes in the symptoms that reflect the body's condition, homeopathy generally regards disease, or disharmony of the body, as a dynamic condition. The therapy does not seek to identify the 'invader' and launch an attack; rather it treats the patient according to the disease and assists in strengthening the body to resist the harmful onslaughts of outside organisms or influences.

Homeopathy also considers that all parts of the body are interdependent and therefore the treatment given takes the entire body as a "whole", rather than concentrating on just any one organ or one particular part of the body. This is known as the ***holistic*** approach. No attempts are made to separate mental illness from physical illness, as these are the symptoms of the one individual. The principle of homeopathic medicine entails the administration of ***potencized*** drugs—that is, pure substances reduced through a special process of dilution to a negligible physical quantity in the dose given to a sick person. When a drug is administered in a pure form, it produces an adverse reaction resulting in the symptoms of sickness in a healthy person. When this same drug is ***potencized***, it has the tremendous power of removing the very symptoms that it created in a pure form. Therefore, it must be understood that certain substances, mercury or arsenic for example, that may prove fatal to a human being in pure form have medicinal qualities in their ***potencized*** forms. The homeopathic ***potencizing*** process, also known as ***dynamisation***, is simplicity personified. One part of a drug is diluted with 99 parts of an inert or non-medical substance, such as sugar or milk or rectified alcohol, through rubbing or shaking until "thorough absorption" is achieved. The resultant total is said to have "potency 1". If one part of this "potency 1" is diluted with another 99 parts of the diluent we get "potency 2" – and so on and so forth. By repeating this process, potencies of 100, 1000, 10,000 and even higher can be achieved. This has a very profound significance because the ***curative power of a drug increases in direct proportion to successive dilutions.*** The other important fact is that the ***reduction of various toxic drugs to such miniscule and negligible physical quantities totally cancels out the otherwise poisonous effects they may have had on the body***. That is why homeopathic medicines are completely non-toxic and do not have any side effects.

Homeopathy and naturopathy both believe in the healing power of nature or the vital force. Both systems, besides being holistic in approach, also lay stress on the importance of fresh air, balanced diet, exercise and peace of mind. The only difference between the two therapies is that homeopathy operates as a systematic science that has a set of rules governing the administration of drugs to sick people. Homeopathy is also often confused with herbalism, because both systems use herbs as medicines. However, there is a vast difference in the methods employed for the preparation of the materials. Whereas herbalism allows the freedom to "mix and match" and there are no set formulas or dosages, homeopathy follows laid down rules, a strict scientific procedure and an exact process for producing medicine which is then prescribed according to the "law of similars". The therapy is cheap because no laboratory testing or extended research is required. Each patient is treated exactly as his or her particular symptoms dictate from the holistic point of view. The system is a proven form of effective medication and the efficacy of homeopathy has been demonstrated by clinical experience for over 150 years.

ACUPUNCTURE. This therapy originated in China over 5000 years ago and involves the puncturing of certain predetermined points on the body with thin, light needles as a means of restoring the body's vital energy and enhancing the working of the immune system in the treatment of diseases and the maintenance of good health. These specific points on the body, barring a few, lie mainly along specific meridians or channels. The therapy was discovered quite accidentally. In ancient wars, the main weapons were swords and spears or bows and arrows. Some soldiers found that when arrows pierced them at particular points not only did the wounds tend to heal speedily but also, in many cases, previously contracted chronic ailments and other medical complications cleared up by themselves. Chinese physicians of that era were quick to latch onto this amazing phenomenon, and through the process of trial and error and patient experimentation, they were able to firmly establish the efficacy of acupuncture as a successful treatment protocol.

According to Chinese philosophy, the cosmos, universe and nature take on the form of ***Yang*** (positive force) and ***Yin*** (negative force) and there is a constant, on-going struggle between these forces. Since mankind forms part and parcel of the universe, all the laws applicable to the universe apply to human beings as well. Therefore, health and disease will be denoted or determined by the fluctuations of ***Yang*** and ***Yin,*** the balance between them and whether the normal vital energy, or ***Chi***, is flowing harmoniously through the meridians or channels. Any excess or deficiency of ***Yang*** and ***Yin***, in whatever form, will cause a dysfunction in ***Chi*** resulting in disease or debility in the individual. The cells of all living organisms being composed of these opposing forces, health depends on a perfect balance existing between the two. Any imbalance will be reflected as sickness or disease. Through acupuncture the energy imbalance can be reinforced or reduced according to need, so that regularity in the performance of ***Chi***, which passes through all the meridians, is achieved and vital energy will continue to correctly regulate the vital functions of the body.

The meridians, or channels, are imaginary longitudes within the body which end at specific vital organs while their superficial path lies close to the skin on which the points to be pricked are located. There are 12 regular bilateral meridians, 15 collaterals, 12 divergent meridians and extra meridians. The points mentioned are "windows" in the channel where maximum energy is infused and therefore, are the most likely spots for effective regulation of vital energy through a needle prick. To date, 750 points have been discovered on the body that can be manipulated to create the required balance through either reinforcement or depletion. ***Chi*** passes through all the organs of the body and Chinese medicine classifies these as **Zang** or solid organs (heart, lungs, kidneys, liver, spleen pericardium,) and **Fu** or hollow organs (stomach, intestines, bladder and the body cavity). The 'extraordinary' **Fu** organs are the brain, gall bladder, bone marrow, blood vessels and uterus.

In 1979, the World Health Organisation accepted and recognised acupuncture as one of the leading tools of therapy and listed about 43 diseases that can be effectively treated by this Chinese method. Since acupuncture adopts a holistic approach and is easily integrated with other traditional medicines as well as different therapies, it possesses a very wide scope in treatment administration. As the system helps in the all round well being of an individual by toning up or relaxing the body as necessary, it goes without saying that it is an ideal vehicle for promoting life, greater vitality and the higher purposes of existence. As good and as efficacious as it may appear, acupuncture does have a few drawbacks, including:

- Faulty techniques and faulty manipulation.
- Possibility of injury to vital organs, tissues, nerves and blood vessels.
- Inability to pierce the proper point.
- Could lead to fainting.
- Needles could break or rust leading to serious infections.
- Could cause bleeding, severe pain, a drop in BP, a drop in blood sugar, abortion or premature delivery, penetration of the joints and ear damage.

- Extra precautions are required at some points which may make the therapy not feasible.
- It cannot tackle an emergency situation.

Acupuncture **CANNOT** be used:

- In malignancy.
- When there are clear indications for surgery.
- On children below six years of age.
- On patients dependent on certain drugs.
- On individuals with weak vitality.
- On an empty stomach.
- On very old patients.
- On patients with a pacemaker or prosthesis.
- Immediately after fasting.
- Immediately after sexual intercourse.
- Immediately after a hot bath.
- Immediately after meals.
- On patients susceptible to hemorrhaging.
- If under the influence of alcohol.
- If sweating profusely.
- During periods of excessive emotion or excitement.
- During pregnancy.

ACUPRESSURE. The concept, approach and philosophy of this therapy are exactly the same as that of acupuncture. It is a simple, safe and effective method of applying pressure on specific points (instead of inserting a needle!) using the fingers in order to achieve a maximum therapeutic effect. Due to certain drawbacks, limitations and technical difficulties in carrying out acupuncture therapy, practitioners who were looking for substitute methods eventually discovered that the various so-called motor points or 'acu' points could be stimulated by other means to get the same therapeutic effects. Trial and error revealed that the points could be stimulated through firm finger pressure and therapeutic benefits heightened if followed by a massage of the affected area. And so, acupressure came to be accepted and practiced as a form of oriental massage. There is absolutely no doubt that, if carried out in the correct manner, acupressure can provide the same healing aspects as acupuncture.

The scope of acupressure is more pronounced than that of needle therapy in some respects. In situations and at body locations where acupuncture cannot and must not be used, acupressure plays an important role as there are absolutely no restrictions on its applications. It has also been noted in some cases where the application of firm pressure is followed by a massage, the patients seem to get better relaxation than that afforded by acupuncture. Also, if the treatment for a specific disease is carried out correctly and methodically, the results in many cases are better than what might have been obtained through needle therapy. Acupressure is the safest form of treatment and can be practiced at home with the leeway of optimum repetitions for obtaining maximum results. Therefore, it is only natural that this therapy is a vital tool during periods of rehabilitation and convalescence.

NATUROPATHY. This is a very scientific system of treatment that incorporates the judicious use of all the methods and procedures of nature cure. Through these methods, health is maintained by relieving suffering, and diseases are totally eliminated without the use of toxic drugs or any injury to vital organs. The history of nature cure almost parallels the origins of man. Ample anthropological evidence points to the fact that this system of therapy was regularly practiced in various forms by many ancient civilizations. Massage, diet and hygiene, physical culture and exercising, and various methods of bathing were some of the many forms of naturopathy employed by the druids of old. Ancient medical manuscripts, collected from all over the world, are unanimous in their praise of and reliance on the virtues of water, sunlight and fresh air and, in the context of curing diseases, their beneficial effects on the health and well-being of humans. Writings by Hippocrates on hydrotherapy confirm the importance and vital role played by naturopathy in the medical therapies of old. In India, the history of nature cure goes even further back and concrete evidence confirms that our ancient sadhus, yogis and sages were the earliest exponents of this system—even before Plato, Aristotle or Hippocrates were born! These ancients routinely subjected themselves to the various rigorous regimens to purify the mind, body and soul as a means of enhancing will power and self-control. Despite its ancient and proven antecedents, naturopathy was only recognised and its practices regenerated about the mid 19th century.

The leading axioms of nature cure philosophy are the ***unity of disease*** and ***unity of cure***. Naturopathy advocates that perfect health means the invisible force, or vital energy, that stimulates the body in health and disease. The maintaining and upkeep of this vital energy through natural means—sleep, sunlight, fresh air, water, mental condition, healthy dietary habits and a regulated lifestyle—is the key to enhancing the body's disease-resisting abilities or raising the efficiency of the immune system. If the body's own immune system is in a position to ably and effectively counter morbid agents, it stands to reason that such individuals will always enjoy good health.

By ***unity of disease*** is meant that if part of the body becomes diseased, then the whole body is affected accordingly. Disease is to be considered as a unit. Any deviation from or violation of nature's laws will encourage the onset of disease which could present itself in a variety of forms and in varying degrees of severity. Naturopathy neither denies nor neglects the concepts of micropathogenic organisms. The truism of nature cure is that these organisms will enter and attack a human body only because that body's vital energy has already been weakened and it has lost its will to fight. By ***unity of cure*** is meant the belief of naturopathy that healing power emanates totally from within the body. The body possesses its own inherent self-curative forces that can both revitalize and revive the debilitated condition of the vital force or energy. Once vitality is reintroduced, the blood purification process scavenges all poisonous and/or toxic substances out of the life-support systems thereby allowing the body to recuperate, cleanse itself internally and regain a state of good health once more. This cycle ensures cumulative beneficial effects throughout the body. The only hurdle erected in the path of this healing and restorative cycle is when the immutable and infallible laws of nature are grossly violated. It is only when these laws are strictly obeyed that the preservation of fitness from within becomes possible and a cure can be instigated to enable removal of the cause of illness, prevent its recurrence and permit the regulation of perfect life and health.

The five elements of nature cure and their therapeutic importance are as follows:

1. **Return to natural principles** by the regulation of eating, drinking, bathing, breathing, resting, working, dressing, thinking, living a normal life with sexual and social activities and establishing them on a moral and natural basis. The therapeutic importance of these habits mainly centres on the maintenance and

upkeep of the body's vital force thereby increasing an individual's disease resisting ability or immunity.

2. **Chemical remedies** which include scientific food selection and combinations, biochemical remedies, simple herb extracts and chemical remedies. The therapeutic importance of these remedies is the assurance of nutritional supplies to the cells and tissues, purification and regulation of blood flow, ridding the body and various systems of toxic wastes and substances and ensuring a balance in the body's essential elements and inorganic salts.

3. **Elementary remedies** centre around allowing the skin to keep breathing thus keeping the pores active, revitalization and relaxation of the body, inducement of sweating and drainage of waste products and the improvement of arthritic and perivascular diseases.

4. **Mechanical remedies** such as corrective gymnastics, massage, magnetic treatment, structural adjustments, cases of surgery and accident. The therapeutic importance of these is the direct and beneficial effects on toning up of the body, improvement in musculature and strength, improvement in blood flow and supply, ensuring the proper function of glands and organs, fast and effective healing of trauma sites, the proper regulation of the circulatory and nervous system, relaxation of ligaments, tendons and vertebral muscles.

5. **Mental and spiritual remedies** such as scientific relaxation, normal suggestions, constructive thoughts and the prayers of faith. The therapeutic importance is the immense improvement in mental faculties, inner strength, purifying the mind, body and soul and increasing will power and self-control.

The scope of treatment through naturopathy actually borders on the limitless. The categorical fields that are being specifically explored for the proper utilization of nature cure are:

- **Psychotherapeutics** in the form of constructive suggestions, correction of habits, psychosomatic treatments etc.

- **Physiotherapeutic treatments** such as articular, spinal, muscular manipulation, massage, vibrations, exercises, rest, orthopaedic gymnastics, hydrotherapy, neuropathy, zonetherapy, sunlight, air, water, earth, cold, light, colour rays, electrotherapy, osteopathy, chiropractic, spondylotherapy, sun and air baths, scientific relaxation, magnetotherapy etc.

- **Chemical and material** therapies involve nutritional control and dietetics, external applications, mineral salts, vitamins, herbs and external applications such as balms, oils, salves, mud packs etc.

Nature cure does not limit itself to any particular 'right' method but utilizes combinations and adaptations of all systems strictly in accordance with the laws of nature by incorporating methodologies of drugless healing. Taking cognizance of all aspects, it may be safely concluded that, on the whole, naturopathy includes all agencies, methods, systems, practices and ideals.

ORIENTAL DIAGNOSIS. There are two forms of this ancient and traditional Chinese diagnostic procedure, mainly Pulse Diagnosis and Tongue Diagnosis. The underlying philosophy for both forms is basically the same, in that, the pulses and the tongue have a direct connection to the body's internal vital organs and both the general condition and the functioning efficiency of these organs can be accurately judged through a physical examination of either the pulses or the tongue. From the traditional Chinese viewpoint, it is important to ascertain the balance of energy in each vital organ. Any imbalance, be it on the excess side or the deficient zone, can indicate the nature and severity of the disease and also the body's ability to cope from the defensive or immunological point of view. Therefore, when a physician is able to ascertain the energy and functional status of an organ, he can also accurately determine the general physical condition of an individual. When the nature of the disturbed equilibrium of vital energy is ascertained, the disease can be easily diagnosed and treatment started immediately. In this type of diagnosis, it is possible to pinpoint a particular disease even before the malady can clinically manifest itself. An early diagnosis facilitates not only early preventive treatment, but also initiation of the appropriate steps to restore the balance of vital energy in the organs. This allows the body to utilize its own internal mechanisms to successfully rid itself of the problem completely.

In ***pulse diagnosis*** the wrists are used as examination points. The wrists of both arms have three superficial and three deep pulses each, the former are felt with light pressure while the latter require deep or strong pressure.

Each one of the internal organs corresponds to one particular pulse. On the left wrist, the superficial pulses connect to the small intestine, gall bladder and urinary bladder while the deep pulses lock on to the heart and pericardium, liver and kidney. The right wrist superficial pulse is associated with the large intestine, stomach and urinary bladder while the deep pulses look after the lungs, spleen and kidney. As per traditional Chinese lore, each pulse has twenty-eight pathological variants for determining whether there is or is not a balance of energy, which may require correction. The imbalance, excess or deficiency, in organ energy is very simply determined by a particular pulse feeling too "full" or too "empty" at a particular position. The frequency, amplitude and quality of the pulse beat are also factors taken into consideration when calculations are being made.

In ***tongue diagnosis*** an observation of the different areas of the tongue and the condition and coloration of its coating will indicate the state of the internal organs. There is a direct relationship between the tongue and its coating and various aspects of the organs such as vital energy, condition of blood and body fluids, channels and collaterals. The dysfunctions of any of these will manifest themselves in tongue indications. Any abnormal changes in the condition of the tongue or its coating is an indication of the nature and changes of diseases from different aspects as well as the pathological condition of different organs and the flow of vital energy. Abnormal conditions of the tongue are indicated by its appearance being pale, red or deep red, purplish, flabby, cracked, thorny, rigid and tremulous or deviated. Abnormal conditions of the coating are indicated by its being peeled, white, yellow, grayish or black.

CHIROPRACTIC. This practice focuses mainly on the scientific manipulation of the spine and is entirely devoted only to the conditional analysis and treatment of the vertebral column. The spine, though primarily constructed of bone, consists of a network of blood vessels and nerves, which supply it with the vital energy without which it cannot exist. This complex network both emanates from and enters into the spine from various parts of the body. Therefore, any interference to any one of these spinal nerves will propagate problems in other areas of the body. For example, if the vertebra is subjected to any sort of displacement or a slipped disc, injury to some nerves will incapacitate their ability to transmit the proper impulses to the rest of the body. This would result in adverse effects on bodily functions becoming the precursor to ill health and the onset of disease. This therapy corrects the mechanical lesions in the body's structures, ligaments and muscles, thereby freeing the nerves and blood vessels of undue pressures.

The therapy also entails focusing a close study of spinal centres as well as the possible effects on peripheral organs and the tissues' stimuli affecting these centres. Stimulation helps in increasing or decreasing the activities of various organs which in turn will excite or inhibit the spinal nerves, thereby contracting or expanding the blood vessels of the concerned organ as the requirement demands. The vertebral construction is composed of various bones attached to each other by ligaments. This forms into a one-piece though highly flexible column that is fitted with special discs in-between acting as cushions. Unfortunately, this flexibility makes the spine extremely prone to weakening. Prolonged improper posturing, injuries, strain and damaged cartilage or ligaments all contribute to weakening the vertebra and inducing improper function. When this happens, there is an adverse interference in the supply from blood vessels and nerves to particular viscera. Advanced age is the main reason for shrinkage of the inter-vertebral spaces. This directly affects the blood vessels, nerves and lymphatic–system of the body, which, in turn, will ultimately affect various organs.

Traction is the main form of spinal manipulation. It stretches the spinal column, opening up the vertebral spaces and thus allowing more nutrients to pass freely into the inter-vertebral discs. The tactile tension applied to vertebrae also stretches the fibres of the spinal discs, thus increasing the spinal circulation, providing nutritional access and correcting any deformities. Traction not only aids in the improved and strengthened transmission of nerve pulses but is also responsible for restoring the forward and backward curve of the spine thus ensuring that the brain and spinal cord remain protected from undue irritation. Stretching also elevates and flexes the ribcage aiding in enhanced respiration. Besides helping to stretch the various muscles of the body—which results in better circulation of nutrients—traction corrects and restores normal posture.

FASTING THERAPY. Fasting entails a complete abstinence from all kinds of food or other energy providing substances—***excepting water and air, which are indispensable***. Fasting instigates the elimination of all toxic and unnecessary products from the system. This therapy also cleanses, purifies and rejuvenates the body finally bringing about natural hunger, which is the main goal. It promotes an entirely natural cleaning out of our "insides" and is a method by which rejuvenation of both mind and body can be achieved. The body is quite capable of sustaining itself for a considerable period of time with internal life-giving nutrition in the absence of food from outside sources. Abstaining from food periodically or fasting during an illness is quite normal and even necessary to physical efficiency because it greatly helps in getting rid of artificial hunger or appetite. Our ostensibly bad eating habits generally result in a plethora of diseases. These are usually due to 1) unbalanced diets, 2) accumulation of unwanted materials and 3) failure of the body's internal digestive and elimination mechanisms to function as they're supposed to. This results in the impairment of the vital energies and life-sustaining functions. Fasting not only rights all these wrongs, but also helps to increase physical and mental energies and increases the power to exercise better control over appetite.

Fasting may be defined as utilizing abstinence from solid and liquid sustenance, except the essentials of water and air, as a means of purification and rejuvenation whereby toxic waste is thoroughly eliminated from the body, thus enabling dead tissues to continue regeneration unimpeded. The body goes through "high" and "low" cycles, as does everything else in the universe, and after some time it tends to build up toxins from chemicals, pollution and overindulgence. When the body starts to cleanse itself, this is a normal detoxification phase and an individual may develop headaches, diarrhea or depression. This is the "low" cycle and the best time to initiate fasting, which would be of enormous help in the detoxification process and in cleaning up the 'insides' more efficiently. Fasting regularly will help the body to heal itself and stay well. There are no particular rules and regulations governing the length of fasting and each case generally depends purely on the individual's physical conditions and needs. A fast may last anywhere from one to ten days. The primary indications for initiating a fast are foul breath, a coated tongue or a bad taste in the mouth. Fasting may be carried out as follows:

(a) A one day fast, at least three times a month.
(b) A three-day fast which helps the body get rid of toxins.
(c) A five-day fast which will start and instigate the healing process.
(d) A ten-day fast that will take care of most problems before they arise.

Fasting for more than three days should be supervised by a qualified health professional. Diabetics should never fast unless supervised by a doctor. Hypoglycemic persons should never fast without also using a quality protein supplement. If fasting is not done according to certain norms, it may bring on the destructive process of starvation, which is extremely harmful. If starvation is not checked in time, it will deprive the body of all nutrition and result in death. Similarly, if coming off long fasts is not done correctly, there is every possibility of causing real damage to the tissues and organs. It is better, for longer lasting beneficial effects, that only raw foods be consumed for two days before commencing and for two days after ending fasting. Eating cooked foods after a fast may ruin the desired effects. Fasting can be ended when any of the following are achieved: 1) hunger is restored; 2) the breath is no longer foul; 3) any coating on the tongue disappears totally; 4) no more bad taste in the mouth; 5) a general feeling of well-being permeates the body accompanied by improved skin texture, brighter eyes and improved vision. Elderly people generally require certain vitamins and minerals daily and they are advised to continue with these while fasting. Nothing should be chewed while on a fast as this action will instigate the digestive process whereby enzymes are secreted into the gastrointestinal tract. With no food in the stomach for these enzymes to digest, trouble occurs. Once the therapy has been undertaken, care must be exercised that the duration of the fast does not deprive the body completely of nutrients. It is beneficial to complete fasting by a preliminary cleaning of the bowels by an enema or a colonic irrigation.

Fasting therapeutically is very beneficial from the point of view of neuropathy because the health of an individual is guided in accordance with nature's laws. All toxic materials and wastes are eliminated from the body which also helps in ridding the cell tissues of accumulated unwanted substances as well as weeding out micro pathogens like bacteria or viruses etc. Thus a curative effect is achieved. If done judiciously and rationally, and according to the laws of nature, fasting not only rejuvenates and revitalizes the body but also increases the effectiveness of the immune system. In this way, fasting will help to promote health and aid in purification of the body, digestive tract and all other cells and tissues. Scientific fasting indirectly helps to accurately diagnose a disease or malady. Regular fasting gives the body a chance to heal itself and rebuild its immune system. Fasting allows the organs to rest and helps in reversing the ageing process. The process cleans the liver and kidneys, purifies the blood, cleanses the colon, helps lose unnecessary weight, gets rid of toxin build up in the tissues, clears the eyes and tongue, cleanses the breath and removes excess water. Ultimately, the physical strength and mental alertness of an individual are enhanced which results in revitalization and an unobstructed flow of vital energy throughout the body for a return to normal health.

PHYSIOTHERAPY. This therapy tones up the body and vital organs through manipulative techniques that are applied completely in accordance with the laws of naturopathy. The three aspects of this form are physical exercise, physical massage and electrotherapy. Diseases present themselves whenever a deviation occurs from the natural way of life. The preservation of health and fitness, in that the body's self-preserving vital energy is working efficiently and safely in purifying our internal dynamism, is a prime accomplishment of physiotherapy. In the physical exercise regimen, regular exercising tones up every physical attribute (muscles, bones, organ functions, blood flow, air circulation, pulse rate, heartbeat, mental capabilities, vision and sharpened senses), which results in the proper functioning of vital glands and organs and the proper internal distribution of nutrients. Physical massage is applied to limbs or parts of the body that suffer impaired function and accomplish the same benefits as exercise by the heat that is created through motion and friction. Massage acts as an artificial stimulation to increase the blood supply in the affected part. In electrotherapy, powerful muscular contractions and vibrations of the voluntary and involuntary muscle fibres occur on the application of electric current, which aids in toning up the nerves and providing heightened and prolonged sensory effects. This in turn helps to clear out any toxic matter that may have resulted because of faulty metabolism. The latest additions to these norms are diathermy and ultrasonics which produce and pass through the skin a succession of pressure waves that penetrate to the muscles and bones causing strong movements in the tissue particulars. The subsequent generation of heat and warmth produces good and desirable results.

All forms of the therapy must be applied scientifically and methodically if any beneficial results are to accrue. Exercise therapy increases, universally, the body's entire circulation so that general musculature is upgraded and the physical form gets stronger to be able to cope with hard work. Exercise also tones up the body, improves defective posture problems and plays a vital role, in accordance with naturopathy laws, in the elimination of toxic materials thus allowing the body to purify itself.

Massage increases internal functional activity, helps supply more nutrition and aids in effective nutrient absorption.

CUPPING. This is a technique, used in China and other Asian countries almost from prehistoric times, of traditional Chinese medicine which employs the application of suction to the diseased areas in order to increase regional circulation and thereby promote healing. The suction, which is applied through a hollow vessel, ceramic, glass or bamboo cups, artificially creates a partial vacuum, which produces blood stains and even blisters at the application sites, thus stimulating the points. Scientific research into this method has revealed that cupping is superior to acupuncture in some cases. It has a wide scope of applications for treatment of low backache, sprains and soft tissue injuries. Cupping eliminates secretions from the lungs and has applications for chronic bronchitis and the umbilical region where acupuncture is forbidden. It supplements acupuncture in the treatment of bronchial asthma, chronic diarrhea and arthritis. The therapy aids in removal of toxic wastes, improves the body's resistance to disease and enhances the immune mechanism. It promotes health for the higher purposes of existence.

IRIDOLOGY. This is a diagnostic science wherein the disorders of the body, in any form, can be detected by examining the iris of the eye. The procedure focuses on the belief that any disease or disorders of any single organ will affect all the organs and induce pathological changes in the entire living body. The abnormalities of the organs are reflected by pigmentation changes in the iris, which are used as a diagnostic index of the disease. This is achieved by noting the presence of different colours or spots, or the possible protrusion or elongation of the structure of the iris. When the normal coloration and the anatomical configuration of the iris are disturbed, depending upon the severity of the disease, the density of the iris tends to change radically. Colour damage to the iris is indicative of functional or metabolic disorders of an organ or a system. Positive changes in the architecture of the iris indicate the presence of foreign matter, infection, drug abuse, intoxication, an acquired illness, poisoning or a wrong treatment. Iridology is especially helpful in diagnosing modern occupational diseases suffered mainly by miners, stonecutters, factory workers and sweepers to name a few.

HYDROTHERAPY. Hydrotherapy is that branch of naturopathy where all forms of water are the main therapeutic agents in the treatment and cure of diseases. Water has been effectively employed from ancient times as a curative agent for most maladies since its main properties are that:

1. It can powerfully absorb, communicate and give up heat.
2. By acting as a universal solvent, it has tremendous solvent properties.
3. It is readily available in the three main forms - solid, liquid and gaseous (steam).
4. It can adapt or be molded to any shape.

Hydrotherapy is the most natural means of cleaning the body and is known to improve circulation as well. Systematic treatment with water acts as a preventive step for warding off diseases as it balances heat gain and loss and keeps circulation in equilibrium. It is a very effective way of eliminating toxic wastes and its curative aspects ensure that the body is kept free of noxious elements thus enabling proper metabolism. Drinking water as a treatment helps get rid of urea, salts, excess sugar, uric acid and many other chemicals that may remain in the blood flow or are ingested with food. It can also be used internally as an enema or colonic irrigation to clean out the bowels. Water cleans out the pores of the skin allowing it to breathe more efficiently by removing dirt as well as dry skin. It is a vital agent for the relief of pain and provides a soothing effect to encourage relaxation. It can be used to excite or depress the nervous system as well as improve blood flow. It can also be used to stimulate the nerve centres, the heart, the muscles, the digestive system and the expansion/contraction of the blood vessels. Water's beneficial aspects are very evident in the treatment of fatigue, anemia, rheumatism, gout, diabetes, sprains, dislocation, peripheral vascular disease and arthritis. Since diseases are prevented by hydrotherapy, another very important aspect of water cure is its positive action in the promotion of health.

Therapeutic Effects of Hot Water. Hot water applications normally prepare the body for cold water applications. Hot baths are normally given to relax and soothe the body and to relieve pain. Hot water can excite or depress the nervous system and the benefits are most evident in a foot bath. The reactive power of hot water plays a vital role in fatigue, rheumatism and the treatment of enfeebled persons. Used as an internal heat application, it is very effective as an enema or in treating rectal irritation. Turkish baths apply hot air at high temperatures without injury.

In this system, dampening is accomplished in the form of steam, radiant heat or hot air. It is a convenient way of inducing perspiration which in turn provides a means of elimination and, therefore, is used to treat gout, toxemia and diabetes. Natural hot springs contain chemicals and minerals in various combinations that are of great therapeutic value. Bathing in hot springs helps to stimulate and treat nerve centres, blood vessels and the heart. In addition, these stimulate and treat all types of muscles, the digestive system, and absolutely all the organs, tissues and cells.

Therapeutic Effects of Cold Water. This directly effects the improvement in the toning up of muscles and blood vessels. Cold water affects the body by contracting the blood vessels, causing a pallor and coldness. As soon as the application is removed, the blood vessels expand and circulation is restored, with the redness of fresh blood flow effectively dispelling the pallor. The application thus first causes a contraction or depressive negative effect that is immediately followed by the beneficial tonic effects of increased circulation. The therapeutic aspects are most pronounced when hip and genital baths are taken.

ELECTROTHERAPY. The scientific use of electric current helps to enhance the body's defence mechanism which in turn enhances the equilibrium in the physical, mental and spiritual fields. The positive effects of current applications on the walls of the blood vessels are the removal of unwanted deposits and the increase of blood circulation in the arteries and veins. This therapy also causes hyperemia (an increase in the blood flow) of an area, increases functional activity and effectively both supplies and absorbs nutrients. It increases both secretion and excretion of wastes through the natural process of elimination. Of the manifold advantages of this therapy probably the most important is its diagnostic value. Current applications enable the gathering of pertinent information regarding the brain, lesions of the spinal chord, muscular degeneration, structural dislocations and malpositions within the body and the location of foreign bodies. Providing both the diagnostic and prognostic aspects of disease, electrotherapy provides invaluable aid to physicians in gauging the extent and planning of effective treatment for any dysfunctions or diseases. Though this therapy is not a cure, it gives valuable remedial help and is of especial use in orthopedic and neurological cases. The convulsive reactions caused by electric current break up deposits in the joints and help to eliminate wastes. By aiding in the stimulation of the organs and body fluids for self-purification, it provides a calming and soothing effect on the nervous system. It also helps to relieve tension, obstructions, undue pressure and any other abnormalities of the tissues, organs, muscles, joints and bones.

PHYSICAL EXERCISE. According to the philosophy of nature cure, disease occurs when there is a deviation from the laws of nature in living and thinking. Health means the harmonious functioning of all the parts that form the entire body. The preservation of health and fitness can only be achieved through the methodical utilization of all forms of exercise, which ensures that the body's vital energy works safely to purify its internal dynamics. Exercising pumps up the body's defensive mechanism to increase the self-curative forces thereby intensifying the self-purification aspects. With this internal healing power working at peak efficiency, the body gets sufficiently toned up to forcefully resist the onset of disease. Exercises done scientifically and properly increase the functional activities of the organs, tone up body musculature and ensure proper elimination of intoxicants in a natural manner, bringing about greater mental and physical output. Exercises play a very vital role in circulatory and musculo-neurological improvements.

Other Prominent Therapies

There are some other prominent therapies that follow similar principles of treatment and cure. However, our discussion here will be limited to only brief descriptions:

- **Aroma Therapy**. Aroma or Essence therapy is a gentle complementary treatment where the essences extracted from plants, shrubs and trees are used for therapeutic purposes.
- **Auriculotherapy.** This is a branch of acupuncture which makes use of the external ear to diagnose as well as treat illness.
- **Bach Flower Remedies.** Bach flower therapy is a specialized system of complementary treatment where medicines derived from specific medicinal plants are advocated for holistic treatment.
- **Bio-Chemic Medicines.** This system of medicine is a specialized branch of therapeutics where specific tissue remedies are used for

the treatment of diseases on the basis of their physio-chemical process in the organism.

- **Electro Homeopathy.** This is a simple, unique, natural and harmless scientific system of therapeutics where the inherent powers or essences of the vegetable kingdom are used on the basis of polarities (positive and negative relations) between the proper remedy and the patient's constitution along with the drug strength and its ration of organic functions.

- **Gem Therapy.** Treatment with gemstones is a recent discovery in the field of complementary medicine where specific gems are used for the treatment of specific ailments.

- **Magnetotherapy.** This is a specialized system of complementary medicine, where natural and artificial magnets are used for therapeutic purposes in order to treat diseases.

- **Moxibustion.** This is a therapeutic method of Chinese medicine that treats diseases by burning "moxa wool" or generating similar forms of heat on or near specific acupuncture points.

- **Neurotherapy.** This is a specialized branch of complementary medicine, which deals with the stimulation of the nervous, circulatory and lymphatic systems in order to give a general tonicity to the body's entire mechanism.

- **Osteopathy.** This is a natural system of healing where the bones are manipulated to help correct misplaced or displaced body parts, organs and tissues.

- **Reflexology (Zone Therapy).** This therapy is a scientific method of treatment where pressures are applied at specific regions on the sole of the foot to cure a specific disease through the systematic stimulation of the reflex zone to achieve therapeutic effects.

- **Vibrotherapy.** This is an effective supplementary treatment which is highly beneficial for the whole body.

But Are the "Alternatives" Really That Great?

No logical argument can refute this fact! There is a widely held view that medical advances in the western world enable us to live longer. This is not necessarily true. The search for all-encompassing cures and the possible methods of prolonging life have exercised the human mind for centuries. In fact, it was actually Man's failure to accept these ills as inevitable which first led to an interest in medicine as a discipline. The average life expectancy at birth during the Middle Ages was quite low, approximately 23 years, and this gradually increased to around 40 years by about 1850. From 1900 onwards, life expectancy continued to soar till today's impressive average figure of around 70 years was eventually reached. However, there's no real reason for us to be overly smug regarding this achievement. While in the last century medical science has succeeded in pushing up life expectancy at birth by some twenty to thirty years, ***the life expectancy of a sixty-five year old during the same period had risen by less than one year!*** While modern medicine has almost eradicated infectious illnesses of childhood and adulthood, ***it has not succeeded in preventing degenerative processes and malignant growths***—the two most common diseases of middle life and old age. In fact, modern medicine has done little to cure or prevent disease when compared with the advantages conferred upon society by good sanitation, smaller families, improved housing and other social improvements!

There's no denying that a handful of drug innovations certainly ***have*** made a difference to life expectancy—the antibiotics, for example. But these were hardly responsible for saving as many millions of lives as one may think, simply because only a mere fraction of the world's population was fortunate enough to have access to them at the time. As a sort of ironic "reverse syndrome", today's greatest concern is the millions of people being ***unnecessarily*** exposed to antibiotics, and to a large extent this applies to other drugs as well. This over-exposure to drugs produces a whole new kind of illness, **iatrogenic illness—*disease actually brought about by doctors!*** Surveys have confirmed that between 3 and 20 percent of all hospitalized patients suffer from side effects caused by their drug treatment. Of much greater concern are those millions of others who are going about their daily lives in an unhealthy state, with drug induced side effects, but are not yet sick enough for hospitalization!

A doctor's duty is to be aware of the problems inherent in frequent changes of ideas and therapies and to protect his/her patients from newfangled drugs, treatments and investigations until they are of proven value. However, in the context of what we have talked about in this chapter, many of these treatments have been used for centuries and have definitely passed the test of time with flying colours. All that is now required is the simple need for them to be evaluated in terms understandable today. We have certainly come a long way from the treatments with plant extracts and simple inorganic compounds of 150 years ago to the antiseptic and safe surgical procedures of today, but we still haven't done much for our survival figures. Modern medicine ***can*** cure some acute diseases and infections and relieve symptoms on a very wide scale, but, at the same time, we continue to live with the great number of unpleasant side effects from drugs and the rampant practice of unnecessary surgical procedures like tonsillectomy, gall-bladder removal and appendectomy—to mention only a few—and the added fact that not all surgery comes into the category of life-saving modern medicine. When someone seeks medical help, he or she is usually suffering from (1) a condition that may be self-limiting; (2) a chronic condition with an uneven succession of partial recoveries and relapses; (3) a condition with a psychological origin or (4) an actually acute medical condition. Whatever the condition, all these people need varying degrees of medical attention and consideration, which, unfortunately, orthodox medicine cannot entirely provide. This is the one situation where the veracity, usefulness and versatility of alternative medicines and therapies have proved themselves time and again with positive results for the total benefit of patients. Indeed, before scoffing at alternative therapies, modern doctors would do well to remember that the very large numbers of people who seek their help generally fall into the first three categories mentioned above—categories which orthodox medicine has very little to offer by way of effective treatment.

In today's world, from a medical point of view, the problem is that people are free to choose how they make themselves ill but NOT how they make themselves well. This is certainly neither an ethical medical approach nor an ideal situation. A patient should not only be given complete knowledge regarding the particular dilemma plus the available treatment options, but should also be free to choose a particular therapy without the fear of being ostracized by the orthodox medical fraternity. Alternative medicine makes no such distinctions because of its holistic approach to the treatment of maladies. Ideally, the world should get the best of both traditional and western medical systems, but this turning point will only come about when alternative medical therapies are considered as first line treatments in their own right. In the meanwhile, alternative medicine will continue to prove a viable therapeutic option without any preconditions.

CHAPTER 12

HEALTH-BENEFICIAL HERBS, VITAMINS AND MINERALS

The Newfangled Angles!

We have given you a semi-comprehensive overview of alternative medicines and therapies in the previous chapter including the reasons for the new-found reliance on traditional remedies. There is no doubt that every single human, besides aspiring to a healthier life, would also like to take the necessary steps to prevent any onset of some of the more life-threatening diseases. The big question is ***HOW?*** At the risk of sounding like a cracked disc, we iterate the urgent and necessary "need to know". You must familiarize yourself with the latest information on health that can be picked up from any bookstore, downloaded from the Internet or available in the regular bulletins published by various health organisations. At the very outset, we specifically mentioned that this book is meant as a means of education—not only to teach you about the functions of your body but also to provide you with information regarding the marvelous "new" therapies that are successfully combating some of the deadlier ailments. In a similar vein, this chapter seeks to deal with remedies and cures that are brought about by some herbal, vitamin and mineral nutritional supplements that have already undergone extensive research and thorough testing. These have proved to be extremely effective in the treatment of various diseases. However, some words of warning. **DO NOT** ***take this to be any sort of guide to self-medication***. Nothing can take the place of traditional medical diagnosis and prescribed treatment. By all means, you are absolutely free to try out some of our suggestions, but only do so ***after having first taken your doctor into your confidence!***

Miracle Cures in Abundance?

That's definitely a fact! The problems and diseases covered here are those that we have spoken about in earlier chapters. These are discussed in alphabetical order, including treatments and remedies, for clarity and easy future reference. The treatments and remedies for each disease include the beneficial herbs, vitamins, minerals and home remedies appropriate for the specific disease. The nutrients mentioned are safe and effective if administered as per the manufacturers' given guidelines, but we still urge you to take your doctor's advice before you decide to embark on a supplement programme. Also, since all of these items are readily available in the open market, you are advised to buy only the products created by reputable manufacturers. At this point we would like to inform our readers that we have successfully tried out certain combinations of herbs, vitamins, minerals and nature cures at our clinic. The results have been extremely encouraging with the treated patients showing positive reactions, without any adverse side effects, to the experiments. We now plan to utilize these additional medications on a regular basis.

Arthritis – No Longer the 'Crippling' Problem?

There are numerous types and forms of this disease, the most common of which are **osteoarthritis,** which is a degenerative joint disease and normally affects people over the age of fifty-five, and **rheumatoid arthritis** which is a type of inflammatory arthritis that attacks the synovial membranes surrounding the lubricating fluid in the joints. Unfortunately, there are neither any simple tests nor any rule of thumb to determine who will and who won't be afflicted. The disease is mainly manifested at the joints of the body and generally is characterized by inflammation, swelling and/or pain. Each joint is provided with its own set of shock absorbers, the ends of the bones having a cap of cartilage, which prevents the bones from grinding against each other. Another layer of cushioning is the moist, flexible, and sinewy connective tissue that holds the bones together. Providing an additional cushion is the synovial fluid, which lubricates the joint. When this natural "padding" deteriorates or becomes inflamed, the joint is prevented from working properly and arthritis occurs. The prime symptom of arthritis is joint discomfort, which may fluctuate from occasional stiffness and a mild pinch of pain to non-stop unbearable and crippling pain, swollen joints and a total loss of mobility. Those with rheumatoid arthritis will additionally experience fatigue, loss of appetite, low fever and anemia. Arthritis is a chronic disease that worsens with time, progressively deforming and ultimately destroying the joints. Besides the two types already mentioned, other forms of arthritis include **gout**, **ankylosing spondylitis** (AS) and **systemic lupus erythematosus** (SLE). ***Gout*** typically attacks the smaller joints of the feet and hands, generally affecting the big toe, and occurs more often in overweight people and those who overindulge in rich food and alcohol.

The ***AS*** form affects certain joints of the spine, generally causing postural deformities and breathing problems. The ***SLE*** form is a malfunction of the body's immune system in that it mimics rheumatoid arthritis (painful and inflamed joints) but is not a crippling disease.

Awareness of the following factors will help in formulating preventive steps, if such a need ever arises:

- Osteoarthritis is increasingly more common past middle age, so the older one gets the more susceptible to contracting the affliction he or she would become. Conversely, rheumatoid arthritis shows up ***before*** mid life, even occurring in young children.
- Most forms of this malady are far more common in women than in men.
- Since women often wear shoes with high heels, which forces the knee joint to compensate for the changed position of the ankle, the number of women suffering from osteoarthritis of the knee is almost double that of men.
- If family members have developed or suffer from arthritis, ***your*** chances of following suit are that much greater.
- Being overweight is definitely a risk factor. The stresses and strains of added weight on joints have definite adverse effects on the proper functioning of cartilage and connective tissue.
- Damage or injury to cartilage and connective tissue normally results in their being unable to replenish themselves as effectively as before.
- People with work routines which entail unusual physical demands on the body, such as athletes and dancers, are more likely to develop arthritis.

TREATMENT AND REMEDIES. The good news is that remedies ***are*** available which help in reversing or significantly slowing down arthritic damage, improving mobility and getting rid of that nagging and debilitating pain.

Glucosamine and Chondroitin for Osteoarthritis. Both of these are common, natural substances which the body produces daily and which are also ingested in small amounts through the food we eat. Both play a major role in the health and repair of the body's joints. ***Glucosamine*** is an amino sugar, which is a building block and must be essentially present if replenishment of the shock-absorbing materials is to effectively take place. It helps in the creation of the second substance ***chondroitin.*** The deterioration process of cartilage renders it dry and brittle and the presence of chondroitin is crucial in fluid loss prevention. This aids in cartilage maintaining sponginess as well as retaining the ability to act as a protective cushion for the bones.

Glucosamine stimulates the production of several joint nourishing substances that are essential for building and maintaining vital cartilage and connective tissue. When taken as a supplement, ***acting as effectively and beneficially as the glucosamine produced within the body***, it speeds up the body's natural repair process. Chondroitin, on the other hand, blocks the action of nutrient-robbing enzymes, thereby retarding the breakdown of cartilage, besides also keeping tissues moist and well nourished. Both of these supplements work effectively and beneficially alone; however, there is a greater and definite synergistic effect when they are taken as a combination–with glucosamine accelerating the repair process and chondroitin slowing down the damage.

Both supplements can be taken orally and are readily available in separate as well as combined forms. A normal, effective dose is 400 to 500 milligrams three times daily; however, since there are many different formulations available, it is best to either follow the manufacturer's instructions or consult with your doctor. What must be kept in mind is that ***glucosamine*** and ***chondroitin*** are **not** analgesics and do not provide instant relief. Since they enhance the repair and damage-prevention processes, it may take as long as six to eight months before any significant, positive effects are evident. Although no serious side effects have been associated with these nutrients, glucosamine is a form of sugar or glucose, so those with diabetes should exercise caution and only take them with a doctor's advice. Salt is another point to consider, and those who have reduced sodium intake because of high blood pressure or for any other reasons, cannot take this supplement in the form of glucosamine sulphate.

Omega-3 for Rheumatoid Arthritis. Since the body cannot manufacture its own essential fatty acids to aid in various bodily functions, these nutrients must be obtained through dietary supplements. ***Omega-3*** oil contains one or more types of essential fatty acids, and is commonly found in most fish oil supplements, the best sources of which are high-fat coldwater fish, such as salmon. Omega-3 has a natural anti-inflammatory effect and is well suited in the treatment of arthritis. It achieves this by reducing the levels of leukotrienes and other natural body tissue substances which have inflammatory effects. It is a truly amazing nutrient, and has the astounding ability to fight disease on several fronts at once. In the case of rheumatoid arthritis, omega-3 oil substantially reduces the dependence on analgesics and in some cases even eliminates their use. Though fish oils are the richest source of omega-3, certain vegetable oils such as flaxseed oil, canola oil and walnut oil, also contain this nutrient.

Capsules containing omega-3 oil are available and an effective dose is between 2500 and 5000 milligrams a day. As with most nutrient supplements, omega-3 works on a long-term basis, taking from three to six months to kick in and then is most effective when taken on an ongoing and daily basis. Those with blood clotting problems should avoid this nutrient as omega-3 is a natural blood thinner, and too much of it can lead to anemia or excessive bleeding from a wound.

Capsaicin Cream. Capsaicin is the active ingredient in cayenne, the powder made from dried chilly peppers. Used as a cream, it been found to drastically reduce pain and boost joint flexibility in arthritic patients. When the ointment is rubbed into the skin over a joint, the blood vessels dilate increasing both the flow of blood and the supply of nutrients to the joint. Not only is capsaicin effective as a pain reducer, it also actively retards the degenerative process itself.

The cream comes in different potencies, and the two most common strengths are .025 percent and .075 percent. Both potencies are just as effective but you are advised to start off with the .025 percent preparation. If this proves to be slow or ineffective, you can then switch to the higher concentrate. Of importance is how often applications are made and not the strength. For best results capsaicin cream should be applied between two to four times a day – the higher the number of applications, the more dramatic and faster the results! There are no side effects worth mentioning besides a sort of burning or tingling of the skin, which is harmless and disappears soon.

Herbs for Arthritis. The following herbs have also been found to benefit the treatment of arthritis: ***alfalfa leaves, black cohosh, brigham tea, celery seed, chaparral leaves, comfrey, devil's claw tea, feverfew, parsley tea, valerian root*** and ***yucca extract. Feverfew*** is especially good for pain and soreness.

Home Remedies. After some research and experimentation, we have come up with a viable home remedy treatment for arthritis which has proved quite successful in some cases treated at our clinic. The three "concoctions" can be easily prepared at home and do not require a doctor's prescription. Absolutely no side effects are associated with these preparations.

1. Prepare a paste with **one teaspoon** of ***honey***, **two teaspoons** of lukewarm ***water*** and **one teaspoon** of ***cinnamon powder***. Thoroughly massage the affected area with this paste **once a day.**

2. Thoroughly mix **two teaspoons** of ***honey*** and **one teaspoon** of ***cinnamon powder*** in **one cup** of warm ***water***. Drink daily **morning** and **evening.**

3. Mix **half a teaspoon** of ***cinnamon powder*** in **one tablespoon** of ***honey*** and take regularly **before** breakfast.

Blood Vessels – Stronger, Younger, Healthier!

Healthy arteries, veins and capillaries are strong yet elastic, are more or less impenetrable and retain their ability to act as smooth conduits for blood flow. Though blood vessels are also subject to ageing, the process invariably goes unnoticed until it's well underway. Besides the ravages of time, our vessels are also susceptible to damage due to diet, lack of exercise and other bad habits. In short, blood vessels do become unhealthy and tend to lose toughness and elasticity, become weak, flaccid and porous and can tear or rupture easily as a result. Sometimes they can also be weakened by temporary illnesses. Poor blood vessel health is a chronic degenerative process, a condition that is irreversible without deliberate effort. Some of the contributing factors of blood vessel fragility are:

- **Age**. This is probably the main cause of blood vessel problems because with ageing cells do not replace themselves as often, thereby resulting in the tissues of blood vessels becoming less resilient.

- **Diseases**. Inflammation, infections and other conditions do play a major role in damaging blood vessels, but the real villain of the piece is ***diabetes*** which really weakens them to the extent that they are unable to carry out the functions of transporting oxygenated blood and other nutrients to the tissues requiring nourishment.

- **Free Radicals**. As you already know, free radicals are the single most damaging molecule that is responsible for causing almost every known degenerative disease by attacking healthy cells and turning them mutant. Free radicals are produced within the body as well as being ingested during normal breathing. Since the cells of blood vessels are particularly susceptible to such attacks, the stage is set for the initiation of atherosclerosis, coronary artery disease, heart disease and stroke.
- **Unhealthy Lifestyles**. There are a number of factors in our daily lives that adversely affect blood vessel health. Smoking, insufficient exercise and eating foods high in saturated fat are events that are totally within our control but are generally ignored. However, other factors such as a polluted atmosphere and unusual job stress are things that we probably have no control over.

Some of the symptoms of damaged or diseased blood vessels can signal the imminence of a heart attack or stroke; therefore, if any of the following are evident, you should consult your doctor without delay.

- Angina.
- Cramps in the legs during exercise.
- Easy bruising.
- Ruptured small capillaries (visible either in the skin or the whites of the eyes).
- Swollen feet, ankles or legs.
- Throbbing or pain in the main artery of the neck during mild exertion.
- Varicose veins.

TREATMENTS & REMEDIES - THE GOOD NEWS. Fortunately, there does exist a nutrient that is not only capable of keeping our blood vessels in a younger and healthier state for a longer period of time, but it is also able to reverse the damage done by diet, age, lack of exercise and other bad habits.

Oligomeric Proanthocyanidin (OPC). OPCs are **antioxidant flavonoids** that are found in a variety of plants. Two sources extremely rich in OPCs are *pine bark extracts* (marketed as **pycnogenols**) and **grape seed extracts**. If one consumes pycnogenols or grape seed extract, OPCs are being taken in each case but in a different mix since the sources are different. Besides being a powerful antioxidant, OPCs primarily act on the blood vessel system in three beneficial ways. They (1) clean out the harmful free radicals which attack vessel cells, (2) have the unique effect of strengthening the specialized tissues of the blood vessels and (3) exert a strong anti-inflammatory effect. Because of the deterioration of **collagen** and **elastin** that makes them tough and resilient, weak or diseased blood vessels become easily penetrable. OPCs bind with the collagens and elastins that make up arteries, veins and capillaries, strengthening and protecting these materials by warding off the attacks of destructive enzymes. OPCs are also "specialists" in the repair of leaky blood vessels.

Treatment of varicose veins, vision problems related to diabetes and cardiovascular problems with OPCs is extremely effective. They can be consumed orally in the form of tablets or capsules and a daily dose of between 50 to 100 milligrams (standardized to 85% active substance) is sufficient to effect the rejuvenation of blood vessels. Since the contents consumed are water soluble, the ingredients cannot be stored by the body and therefore, continued use is absolutely necessary if one is to derive the total benefits from these nutrients. Human tolerance to OPCs has been pretty well established; however, if one experiences any discomfort or outbreak of rash, use should be discontinued.

Carnitine. This amino acid is essential to heart health and benefits the heart by improving its blood supply and strengthening the body's entire blood vessel system. It is especially useful in relieving the severe leg cramps during walking and exertion that are brought on as a result of atherosclerosis. **Carnitine** is administered intravenously in doses of 500 milligrams.

Ginkgo Biloba Extract. Made from the green leaves of the **ginkgo** tree, this extract is a proven vasodilator in that, it helps the arteries and capillaries to expand thus further enhancing the flow of blood which nourishes every part of the body. The extract is especially effective for the arteries feeding the brain and also relieves leg cramps that accompany impaired circulation.

Ginkgo comes in various forms, including tea. The capsule form of ginkgo is the most effective as this provides a sufficient amount of the nutrient's active ingredients. The recommended dosage is 120 milligrams per day.

Cancer - Still Deadly but Negotiable

We have already discussed cancer at length in Chapter 10, so you may want to flip back for pertinent references. To refresh your memory somewhat, we'll remind you here of the main categories of cancer, some of the main risk factors involved and some obvious symptoms that should trigger both your curiosity and alarm! As you know by now, cancerous tumors have no nerve endings and may not be noticed or felt until they start to push against nearby tissues. Cancers like leukemia are also likely to be well established in the body before they are eventually detected. Even when they are noticeable, each cancer has different and particular sets of symptoms. There are over a hundred varieties of cancer, all having different causes and varying in the speed with which each one spreads. Undiagnosed and untreated, most cancers result in pain, suffering and an early death. However, the advancement of medicine ***has*** prevented cancer from really running amok. The best approach to cancer prevention is to use knowledge about nutrition and nutritional supplements to cut down the risks as much as possible. The protective power (against the onset and spread of cancer) of the nutrients discussed here is not an alternative to regular medical check-ups or a replacement for established cancer therapy. They must be used ***in addition to*** appropriate care and treatment. One fact is undeniable—***if cancer is detected in time and treated appropriately, it is curable.***

Cancer falls broadly under four main categories which cover the entire spectrum of over one hundred varieties of the disease. These are:

1. **Carcinomas** which affect the skin, mucous membranes, glands and other organs.
2. **Leukemias** which are blood cancers.
3. **Sarcomas** which affect muscles, connective tissues and bones.
4. **Lymphomas** which affect the lymphatic system.

There are almost as many alleged risk factors for cancer as there are forms of cancer. Some of them go from the ridiculous to the sublime, but so many of the substantiated ones remain that it is prudent to accept the fact that most of us have some degree of risk. Therefore, the most common risk factors, previously discussed in Chapter 10, are repeated here.

- **Smoking** is an activity akin to building a private toxic waste dump inside the chest, right next to the lungs where maximum damage is guaranteed.
- **Alcohol** in excessive amounts is only inviting trouble.
- **Dietary habits** can literally hurt as well as help. Animal proteins, junk foods, smoked or cured red meats, processed refined foods and saturated fats are dangers to be avoided.
- **Obesity and lack of exercise** are two interrelated factors that are likely to be the prime future cause of cancer. Fat cells are estrogen-rich and excess estrogen is a factor in breast, ovarian and other common cancers. Exercise can reduce obesity and, more than being just a means of weight control, if carried out regularly, can even help reduce incidences of cancer in those who aren't obese.
- **Genetic inheritance** is an overestimated cause. However, inheriting a faulty gene could be a reason, although it's not entirely correct that a single defective gene causes cancer. Defective genes may instigate runaway cell production but healthy genes do compensate for any mismatch. It requires at least a two-stage process for cancer to really be able to take hold.
- **Environmental and occupational exposure** to high concentrations of certain chemicals and toxins and to natural factors such as sunlight can promote cancer.

As mentioned earlier, each cancer has a different set of symptoms and it won't be possible to cover the entire lot here. The indications given below are general in nature, extremely common and the causes are usually not cancer. Notwithstanding, since cancer is a serious and deadly malaise ***and*** really effective treatments are available today to check its spread, a little care and circumspection will pay rich dividends in the long run. Signs to watch out for are:

- Any unusual lump or thickening or tenderness anywhere in the body, particularly in the areas of the breasts and scrotum; tumor or lump under the skin resembling a wart or an ulceration that never heals; moles that change colour or size; flat sores; lesions that look like moles.
- A persistent cough, hoarse throat, sputum with blood or chest pain.

- Unusual bleeding or discharge; blood in the stool; blood in the urine; weak or interrupted urine flow.
- Unexplained changes in vision, balance and coordination.

TREATMENTS – SOMETHING TO SMILE ABOUT! Our body cells are exposed to free radical activity every second of our lives. These harmful chemical molecules enter our systems through the pores of the skin, through the air we breathe, as by-products of food and through a myriad of other means. They cause healthy cells to mutate which eventually results in cancer. Chapter 4 discussed in detail the impact of free radicals on our health. One of the main methods of neutralizing free radicals is the powerful actions of antioxidants on their activity. Though there are several nutrients that act as antioxidants, there are hundreds of types of free radicals and no single nutrient is effective against all of them. However, there are various combinations of nutrients that enhance the antioxidant process of checking the spread of cancer. We shall now talk about some of these.

Selenium. This is one of the trace elements essential to health. **Selenium** forms enzymes that remove two really harmful free radical substances, **lipids** and **hydro peroxides**, from the body. Without selenium our body's defence mechanism doesn't function efficiently because the particular required enzymes do not form. Low levels of selenium in the body mean a low immune response. It is an established fact that Japan and Iceland have low rates of cancer because their soils are rich in iodine and selenium. Finland has created safeguards by adding selenium to soil fertilizer, flour, salt and many other edible items. There is a definite connection between selenium-depleted soil and the high rates of cancer in certain parts of the world. It has been proved beyond doubt that selenium supplements can ward off many types of cancer (breast, lung, liver, gastrointestinal, colorectal, urogenital, prostate, ovarian, cervical) besides providing the additional benefit of preventing heart disease. The same lipids and peroxides that damage the cells are also responsible for hardening of the arteries.

Selenium is available in both tablet and capsule form. There is no real established dosage, but 200 micrograms a day, considered the maximum non-toxic daily dose, is effective in cancer prevention. Experts advise that the best way to take selenium is in organic preparations which are better absorbed by the body. Examples are yeast or garlic that has been specially grown in selenium-rich soil. Selenium by itself provides a fairly effective cancer shield; however, when it is taken along with Vitamin E the buffer to cancer is almost doubly effective. Selenium is safe when taken in the recommended dose and no side effects have so far been reported.

Ginseng. This herb from the **ginseng plant** is grown almost all over the world and is available in many varieties, but we refer here to **red panax ginseng**, which is mainly an Asian product. Ginseng acts by decreasing the number of tumors formed in response to carcinogens and slows the growth rate of any tumors that do develop. It does this by increasing the activity of **NK cells**, which are specialized cells whose prime job is to restrict tumor growth. It is also a powerful antioxidant that neutralizes specific cancer-promoting chemicals. There are certain types of ginseng that are far more effective than others, but the red panax type is probably the king of the crop in the protection against cancer. The use of ginseng in various remedies is recorded in ancient Chinese and other oriental medical texts. This herb is also very effective for increasing sexual function and fertility in men.

Ginseng comes in various forms including tea, bottled tonics and capsules. The best form is the standardized capsule from reputable manufacturers from which you can fairly reliably obtain the guaranteed amount of the herb's active ingredients. Since the root is quite expensive and the strengths vary from product to product, the best dosage would be to follow the package directions. For best results, ginseng has to be taken on a long-term basis and should not be consumed with stimulants such as the caffeine present in coffee, tea or cola beverages. The herb is quite safe when taken in moderate doses; however, certain individuals could be allergic. If one develops heart palpitations or chest pain, nausea, headache, agitation or insomnia use should be stopped at once. The herb can be very toxic if taken in large amounts, so care should be exercised that the label instruction dosages are followed meticulously. One is also cautioned against using ginseng if antipsychotic or antidepressant medications are being taken or if one is under hormone treatment.

Vitamins C and E. These are powerful antioxidants in the body which also effectively neutralize free radical activity. They are being described as a combination not because they have the same role to play in defeating cancer, but because they work with greater impact when used together! Another reason for them to be taken together is that if the **vitamin E** molecule gets damaged by free radicals, **vitamin C** acts as the "transformer" in getting ***E*** back in shape! Both vitamins protect against the cell mutations that give rise to cancer and together sweep up the various free radicals that trigger specific types of the disease. Vitamin C reinvigorates cells damaged by X-rays, besides having a particular neutralizing effect on a free radical molecule which attacks the stomach. When treated with vitamin C, certain types of cancer cells are not only killed but their runaway growth is also blocked. Cervical cancer is known to be brought on by low levels of vitamin E. When used with **beta-carotene**, vitamin E has a protective effect on the cells of the mouth. Although both these vitamins have always been part of the human diet, our current eating habits ensure that we do not get a sufficient supply. Since the efficacy of both vitamins as cancer fighters is indeed a proven fact, all that one really needs to do is to enhance their build up in the body either through supplements or through improved diet.

Vitamin C is available in tablet form, including chewable ones, while vitamin E can be had through tablets, capsules and drops. A point to be noted is that they have a special synergistic effect and should always be taken together. *Effective daily dosages are* **2000 milligrams** of *vitamin C* and **between 200 and 400 IUs** *of vitamin E.* On the aspect of side effects, both vitamins can cause diarrhea and stomach upset.

Flavonoids. These are a vast group of vitamin-like substances basically found in plants. They are also referred to as **bioflavonoids**. The number of varieties available is so large that it is impossible to list all of them. Flavonoids do regulate the body's use of vitamin C, enhancing its beneficial effect, but they also possess specific powers of their own as well. Taken as a whole, flavonoids are powerful antioxidants. However, certain types are known to be especially potent against specific types of cancer, but only a few of these have been identified. Flavonoids fight cancer in two ways. On the one hand they protect us from this disease and on the other they help in recovery. They act as a sort of natural chemotherapy by dramatically slowing the proliferation and production of cancer cells. The big advantage here is that flavonoids don't harm normal cells, their only target being the malignant ones. Studies have revealed that dietary flavonoids are directly responsible for providing the cancer-protection shield; however, the use of flavonoid supplements is quite recent so we may have to wait a while for concrete information. Rather than referring to the flavonoids themselves, it has become a sort of general human tendency to refer instead to the substances that contain them. So when people talk of the benefits of regularly having **red wine**,

green tea, grape seed extract and **pycnogenols**, they are actually pointing at the rich flavonoids that they contain.

A few of the more effective cancer-fighting flavonoids are:

- **Quercetin.** One of the first to gain recognition, this flavonoid is found in common food items such as onions, broccoli and apples. It is particularly effective against lung cancer.

- **Gallate compounds of green tea**. ***Gallates*** are also found in grape seed extracts, but green tea or green tea extract is definitely their richest source. More than 30% of green tea's flavonoids are ***gallate compounds***. These are especially effective against cancer in women.

- **Genistein and daidzein in soy products**. "The proof of the pudding is in the eating" is the reason why people who eat a lot of tofu and other soy products have the lowest rates of some cancers. These nutrients protect against prostate and breast cancers.

- **Tangeretin found in citrus**. Though the vitamin C from citrus fruits seems to get all the credit for health benefits, the ***flavonoid tangeretin*** requires its share of applause for protection against cancer in general.

Flavonoids come in tablets as well as capsules and are labeled either ***quercetin*** or ***genistein*** or are named by the plants from which they are derived. There is no recommended daily dosage for flavonoids. Although these nutrients do not appear to be toxic and there are no reported side effects to date, it would be prudent to follow the manufacturers' usage instructions. Since most of these are plant extracts, there is the possibility of an allergic reaction.

Carotenoids. These are the colour-giving pigments found in red, yellow and orange coloured fruits and vegetables. Though there are over six hundred known **carotenoids,** not all of them are equally useful. About thirty to fifty of them are converted into **vitamin A**, an anti-cancer vitamin, in the body and just a few of the remainder have particular cancer-fighting abilities of their own. Similar to the flavonoids, carotenoids, firstly, act as antioxidants and clean out various cancer-promoting free radicals. Secondly, they have the ability to slow or stop the reproduction of cancer cells. This is an important facet because one of the reasons why cancer is so deadly is the rapid and uncontrolled multiplication of cells. Diets rich in carotenoids are known to lower the overall rates of cancer as well as lowering the rates of specific cancers such as **breast, colon, lung, mouth, prostate, stomach and thyroid**. ***Alpha-carotene, beta-carotene, lycopene, lutein, cryptoxanthin*** and ***zeaxanthin*** are carotenoids that positively act against the spectrum of free radicals known to be associated with cancer. The two most well-known of these nutrients are ***beta-carotene*** and ***lycopene.*** The latter is an example of the fact that carotenoids need not be eaten in a fresh state. While fresh tomatoes contain ***lycopene*** in small quantities, cooked and concentrated tomato products (purees, pastes and sauces)—the more concentrated the better—are very rich in the substance.

Carotenoids are available as tablets and capsules. These can be bought as individual carotenoids, such as ***beta-carotene*** or ***lycopene***, or as a blend of carotenes. The best types are those that are derived from natural sources such as plants. To maximize protection, carotenoids should form part of an overall supplements programme that includes other antioxidants like ***selenium*** and ***vitamins C and E***. There is no recommended effective dose as this will depend on the preparation that is purchased, so it would be best to follow the package instructions. All these are considered non-toxic when taken in moderate amounts and no side effects have been reported.

A few other sources affording protection against cancer are:

- **Vitamin A**. This vitamin is normally formed within the body by various carotenoids, but can be had from other sources as well. Besides being important for overall health, vitamin A has a protective effect against cancer and, like several other nutrients, may play an important part in putting an end to tumor growth. It can be toxic in large doses and the recommended allowance is 5000 IUs per day.
- **Calcium plus Vitamin D**. The combination of ***calcium*** and ***vitamin D*** has proved to be the biggest protector against rectal cancer in women.
- **Coenzyme Q-10**. This substance, found in small amounts in some foods and produced within the body, has already been found to dramatically affect heart health. Its applications for cancer are fairly new and still under study. There are possibilities that it will be useful in shrinking tumors of the breast and liver.
- **Probiotics.** These are friendly bacteria that protect the digestive tract from its numerous harmful cousins. For a variety of reasons,

there are many who just don't have enough of these helpful micro-organisms in their bodies. However, they are now available as supplements in capsule form. Besides aiding in combating intestinal infections, ***probiotics*** are also known to effectively fight against colon cancer.

- **Silymarin**. This is the active substance of milk thistle. In tests it has been shown to enhance sun protectors and therefore aid in the prevention of skin cancer.
- **Alpha Lipoic Acid**. This nutrient is actually good for the body besides being a powerful antioxidant. It neutralizes free radicals and protects against cancer. It also protects other antioxidants, like vitamins C and E and coenzyme Q-10, from harm.

Herbal Aids. The following herbs should be included in cancer prevention or cancer therapy programs: **black radish, dandelion, echinacea, Jason Winters tea, pau d'arco, red clover, suma** and **carnivora** (a South American plant). Many people with external cancers respond well to poultices made from **comfrey, pau d'arco, ragwort** and **wood sage.**

Home Remedies. At our clinic we have developed an herbal concoction, in the form of tea, which has proved beneficial to cancer sufferers in general. In some cases the cancer has regressed completely. The recipe is given below:

Ingredients:

Burdock root (cut)	1/2 cup
Sheep sorrel (powdered)	3/8 cup
Slippery elm bark (powdered)	2 tablespoons
Turkey rhubarb (powdered)	1 teaspoon

Preparation Method:

Mix all of the above quantities into ***2 gallons*** of water and boil the mixture for ***10 minutes***. Remove from heat and allow to cool for ***12 hours***. After cooling, reheat the decoction again and immediately remove from heat the moment it starts to boil. Cool, strain and store in amber coloured, sterilized bottles ***that do not permit light to filter through***. The mixture will lose its potency and effectiveness if exposed to light.

Dosages:

For cancer patients: Make a tea by diluting one fluid ounce

(30 ml) of the decoction with two fluid ounces (60 ml) of hot water. *Sip this herbal tea slowly once daily* ***at bedtime*** or *at least* ***two*** *hours after eating.* No food should be consumed within one hour of having drunk the tea.

As a preventive measure: A daily dose of half a fluid ounce (15 ml) diluted in two fluid ounces (60 ml) of hot water.

Cholesterol and Triglycerides – 'Good' When Low?

These "inseparable companions" have been dealt with at length in earlier chapters, so detailed descriptions are not necessary here. When one thinks about it, it's almost as if the components in our blood, plus those in the food we consume, are sometimes "hell bent" on making life as miserable as possible for us, and we have to just grin and bear it! Recapping a bit, you will remember that **HDLs (high density lipoproteins)** are the good or *Happy* guys and **LDLs (low density lipoproteins)** the bad or *Lousy* bunch in the drama of **cholesterol.** HDL is cycled through the liver and removed from the body while LDL hangs around in the blood, becoming a prime target for free radical attack, and eventually turning into the plaque which causes arterial blockages. Playing footsy with LDLs in this macabre tango are **triglycerides,** which also accumulate in the blood due to improper metabolism. Both substances are manufactured in the body and are essential for normal bodily functions; however, higher than normal levels of cholesterol and triglycerides in the blood mean that you're in for a rough roller-coaster ride health-wise.

If these high levels remain unchecked or untreated, you're a prime candidate for coronary artery disease, heart disease, cardiovascular problems, heart attack and stroke – *lock, stock and barrel!*

Unfortunately, there are no direct early symptoms to indicate the presence of high levels of cholesterol and triglycerides, when the problem could possibly be nipped in the bud. By the time symptoms (angina pains, high BP, breathlessness etc.) start to manifest themselves, you've probably already been subject to the malady of elevated levels for years. Another point, *high cholesterol and high triglycerides don't just go away by themselves!* A concerted and radical lifestyle change on your part (diet, exercising, giving up bad habits, correct and judicious medication, etc.) is necessary if these are to be effectively controlled. However, you'll find that the effort does pay a rich dividend—at least as far as remaining in good health is concerned. Many factors contribute to high *cholesterol* and *triglyceride* levels of which some are controllable and some are not. The risk factors associated with elevated levels include:

- **Diet**. The levels of both cholesterol and triglycerides are influenced by what one eats. Diets rich in meat, animal products and saturated fats will raise cholesterol. Diets high in carbohydrates and low in proteins will raise triglyceride levels.
- **Family History**. Some bodies tend to overproduce cholesterol, and high cholesterol appears to have a genetic link. As a result, and also because bad eating habits are also passed along, problems of high cholesterol often run in families.
- **Menopause.** Prior to menopause, women's cholesterol levels are usually lower than those of men of the same age. After menopause situations become almost equal, with women tending to have slightly higher levels of both cholesterol and triglycerides.
- **Age.** As age increases, there is an upward trend of cholesterol and triglyceride levels in both sexes.
- **Overweight.** Being overweight definitely raises the levels of both substances, but with weight reduction, both levels come down as well.
- **Alcohol Consumption**. Raised levels of both cholesterol and triglycerides will result when there is a high intake of alcohol.
- **Exercise**. There are no doubts that exercising helps raise the levels of HDLs while reducing those of LDLs.
- **Disease.** Certain diseases are responsible for raising cholesterol and triglyceride levels, of which diabetes and hypothyroidism are two common maladies.

And there are several other disorders that can also play a role.

TREATMENTS – SO, IS THE OUTLOOK GRIM? Not on your life! The nutrients that we're going to talk about will bring a sea change for the better. However, some hard work is also required on your part. If your lifestyle undergoes some essential changes and you work on correcting the risk factors highlighted, there's no way in which you're going to lose this battle.

Vitamin C. The strengthening and protective effect that it has on cardiovascular tissues is the main heart-helping activity of vitamin C.

Intake of vitamin C also lowers the iron concentration in the blood. This vitamin's primary activity is to strengthen and protect the heart but it also has the secondary benefit of helping to control serum levels of cholesterol and triglycerides. It is available in tablet form and the recommended effective dose is 2000 milligrams per day. This dosage may be considered too high in some cases and may result in some unpleasant side effects such as stomach upset and diarrhea. If you're interested in taking large doses of this vitamin, it would be best to start with small doses and gradually, over a period of time, build up to the required daily intake. It would be best to consult with your doctor in this respect. Vitamin C will prove effective only if taken regularly on an ongoing basis. That is the only way of harnessing its long term benefits.

Vitamin E. This vitamin is a powerful antioxidant and, in combination with ***vitamin C***, is essential to cardiovascular health. Though most of us have a sufficient ***vitamin E*** intake, it has been established that ***larger amounts are the key factor in lowering cholesterol and preventing arterial damage***. You're already familiar with what free radicals are, the dangers they pose to overall health and the havoc that they're capable of wreaking on our innards. Suffice it to add here that, of the numerous different types of this debilitating molecule some are more dangerous to certain body systems and organs than others. When free radicals bind with *LDL cholesterol and oxidize it,* the new material formed as a result attacks the artery's **endothelial cells**—those specialized cells that allow the artery to expand to accommodate bits of debris in the blood. This material has a penchant for becoming embedded in artery walls, turning into plaque and, over a period of time, narrowing the passageway. The artery becomes more scarred and rigid as a result.

Vitamin E stops this process in the initial stages itself by binding with the free radicals and rendering them inert. All antioxidants, including *vitamin C*, act like this, but *vitamin E* is drawn like a magnet toward those free radicals that particularly affect cholesterol. With the free radical chain reaction neutralized, the arteries are kept free of junk and the endothelial cells are protected against damage. In addition, vitamin E is a natural blood thinner and anticoagulant and also eliminates those solid particles which are not a result of free radical activity.

Vitamin E supplements are available as capsules, tablets and drops. Studies have indicated that, when taken by itself, a daily dosage of **400 IUs** of vitamin E reduces cholesterol levels. However, it is best taken along with vitamin C because when a vitamin E molecule is damaged on interaction with free radicals, vitamin C converts it back to its original form. Also, if taken with **1500 milligrams** *of vitamin C*, the daily dosage of *vitamin E* can be reduced to just **100 IUs.** Some overt side effects of vitamin E are that it can cause nausea, diarrhea and stomach discomfort in certain people. Taking it with food and water will help. Patients with anemia, liver disease, an overactive thyroid or poorly clotting blood should avoid taking the supplements until they have consulted with medical specialists.

Tocotrienol E. This recently-discovered vitamin E molecule promises to be even better than the regular vitamin E in reducing cholesterol and triglyceride levels. There are eight different forms of vitamin E, four are known as **tocopherols** and four as **tocotrienols**. Each form differs slightly in molecular structure. The most commonly used form of E supplements is **alpha-tocopherol**. Though *tocopherols* are powerful antioxidants, *tocotrienols* have proved both far more potent and effective for reduction of cholesterol and prevention of blood clots. Tocotrienol E is known to deliver 60 times the antioxidant force of regular vitamin E. This particular E molecule excels at traveling through the bloodstream by attaching itself directly to the LDL cholesterol molecules and inhibiting their oxidation by free radicals. Tocotrienol E is also an effective blood thinner, but totally without the side effects of drugs like aspirin. It is also an E molecule that provides help in reversing three major threats to the heart: high cholesterol, abnormal clotting and free radical damage.

Omega-3. Eating too much fat is certainly bad for the body, but fats known as **essential fatty acids (EFAs)** are actually necessary for the body. Of special interest are two EFAs known as **omega-3** and **omega-6**. The two *omega-3* fatty acids that are particularly effective in keeping triglycerides low and cholesterol in balance are **EPA (eicosapentaenoic acid)** and **DHA (docosahexaenoic acid)**. Certain cold water fish are rich in omega-3s and as a result fish oil has become a popular supplement. There is a definite association between fish products rich in **EPA** and **DHA** and ***low blood triglyceride levels***, especially when these substances are added in supplement form. Research has positively established that *omega-3* outperforms all other medication/mediums in the ***drastic reduction of triglycerides, LDLs and total serum cholesterol.*** It also acts as a kind of natural blood thinner and keeps platelets from clumping together and forming dangerous blockages. It also strengthens the heart muscle thereby permitting more forceful contractions. Therefore, omega-3 works in many ways to guard against heart attack and stroke.

Omega-3 contained in fish oil comes in capsules as does the ***omega-3*** that is extracted from flax seed oil. Though omega-3 is available from sources other than fish, ***the richest concentration of* EPA *and* DHA *is only in fish oil.*** Since the mix and potency of omega-3 oils vary from product to product, the best method of ensuring that you consume the appropriate dosages is to follow the manufacturer's directions for use. Omega-3 is considered generally safe, but it may cause nausea in some. Patients taking blood thinning medications or those with poorly clotting blood should not take these supplements without first consulting a doctor.

The omega-3 obtained from **flax seed** is very efficient in reducing high levels of cholesterol and triglycerides and provides essential fatty acids that your body cannot produce on its own. In addition, this is a rich source of alpha lipoic acid (ALA) and lignans. ALA helps reduce blood sugar and insulin levels while lignans have anti-mitotic actions, which means that they regulate cell reproduction. Flax seed protects the health of the colon and keeps the bowels cleansed by improving digestion and elimination. It also assists memory function and supports the body's inflammation response. The most effective way to eat this beneficial food is to ***grind it fresh and add it to a cereal or fruit so that the oils and fibres are easily available and absorbable.***

Garlic. This herb is normally used as a delicious flavoring for food but it also happens to have some amazing health benefits. Garlic affords protection against heart attack and stroke by cleaning out cholesterol and other blockage-creating materials from the blood. Taken either as fresh cloves or in standardized preparations, garlic has been proven to:

- Lower overall serum cholesterol
- Increase HDL levels and at the same time decrease levels of LDL
- Act as a natural blood thinner and eliminate clotting by preventing platelets from clumping together
- Perform as a powerful antioxidant

If eaten fresh, the amount required to be consumed is quite large if all the beneficial aspects are to be effectively obtained. However, these benefits are also available in a tasteless and odorless supplement. The substance responsible for the health benefits is **allicin**, but garlic does not contain *allicin!* However, when two other substances in garlic, **allin** and **allinase**, interact, allicin is formed. Unfortunately, allicin has a very short life and, therefore, it is imperative that the garlic supplements be formulated in such a manner that allin and allinase combine and release allicin precisely during the digestive period. If this formulation is not concocted correctly, the supplement will be inactive as well as ineffective. A diet of garlic supplements combined with essential fatty acids not only lowers cholesterol but also has the healing effect of bringing damaged arteries back to normal. In addition, garlic inhibits the release of a body chemical that constricts blood vessels and can also reduce hypertension. Since it effectively fights microbes, garlic is useful for treating infections, colds and flu.

Garlic can be taken fresh and is also available in both tablet and capsule form. The research findings indicate that an effective dose is 900 milligrams a day. For it to be able to work properly, garlic needs to be taken continuously. Cholesterol levels were shown to increase the moment it was withdrawn. Regarding side effects, though garlic is quite safe, some may experience a burning sensation in the mouth, throat or intestinal tract. Stomach upsets and bouts of diarrhea are also known to occur. Patients on blood thinners or with poorly clotting blood should avoid garlic.

Pantethine. This derivative of panthotenic acid (vitamin B5) helps to lower cholesterol and triglyceride levels. Taking 900 milligrams of pantethine a day regularly over a period of three to four months will significantly reduce both total cholesterol and LDL cholesterol levels, while at the same time improving the LDL to HDL ratio. Five hundred milligrams a day has also proved to be an effective dose in certain people.

Guggul. This is a nutrient, also known as **gugulipid**, whose products are derived from the *mukul tree,* which is native to India and is a close relative of the tree from which myrrh is obtained. **Guggul** is widely used for lowering cholesterol and triglycerides and a 500-milligram daily dose is very effective. When guggul is combined with dietary changes, the results are even more impressive.

Niacin. The eight week cholesterol cure! Take 500 milligrams sustained release tablets once a day for a week. Increase the dosage to 1000 milligrams a day for two weeks and further increase the dosage to 1500 milligrams per day (i.e. 500 mg *three* times a day).

Herbs. **Hawthorn** is a thorny, tree-sized shrub of European descent and is possibly the best heart-strengthening herb available today. Its leaves, flowers, berries and bark are all used for herbal preparations, though the highest potency comes from the leaves and flowers. Hawthorn plays a major role in lowering blood pressure besides also lowering the level of lipids in the blood—those insoluble fats that contribute to cholesterol. In effect, it lowers LDL cholesterol as well as triglycerides. As a supplement, the herb is available in tablets and drops. However, ensure that the preparation you buy has been made from leaves and flowers only. There is no recommended dosage and it would be best to follow the label instructions which will differ for prevention and treatment. Other herbs that help in lowering cholesterol are **cayenne, goldenseal** and **kelp**.

Home Remedy. Prepare **16 ounces** of ***black tea*** and mix in **three tablespoons** of ***cinnamon powder*** and **two tablespoons** of ***honey***. Strain and divide the mixture into three equal portions. Drink one portion at a time, either hot or cold, **three times a day**. Take your cholesterol readings every week to check on how much the readings have lowered. Continue the treatment till you are satisfied.

Diabetic Damage – A 'Contendable' Problem?

Diabetes mellitus and diabetes insipidus have both been discussed in detail in Chapter 10 and you already know of the many health complications the disease can cause. Regardless of whether it's Type I or Type II, ***diabetes mellitus*** is a serious disorder requiring immediate medical management. Besides being a complex disease, it is also a malady whose chemistry changes from day to day. Type I is the more serious form of the disorder requiring lifelong medication. Type II is far more responsive to intervention and the condition can be eliminated or greatly improved through weight loss, diet monitoring and some lifestyle changes.

TREATMENTS & REMEDIES – UNDO DAMAGE? IMPOSSIBLE! The good news is that research has adopted a whole new approach to diabetes, with focus centering on the damage it does to the body instead of looking only at insulin problems. This damage, which is caused by free radicals, has much more far-reaching consequences than the lack of insulin. Insulin can always be supplemented, but can the damage be checked or undone? The answer is an unequivocal *YES!* However, the first and most important aspect is to determine whether or not one actually has diabetes, diagnosis actually being the first line of defence. As mentioned in earlier chapters, diabetes is a "silent stalker" and it may be years before truly overt symptoms begin to present themselves. This is especially true for

Type II diabetes. So, if you encounter some or all of the following symptoms, whether benign or severe, consult your doctor immediately for a complete physical and the pertinent blood and urine tests.

- Excessive thirst
- Unexplained weight loss
- Frequent urination
- Periodic bouts of blurred vision
- Extreme and persistent fatigue even after sufficient sleep and rest

Obesity is common among Type II diabetics, so if a person is overweight, there is every likelihood that his/her blood sugar levels will be elevated. The treatments outlined in the following paragraphs are not an alternative to professional medical management. Always consult your physician when taking any of the following.

OPCs. You've already met these versatile characters in the foregoing section on blood vessels. So it makes perfect sense that these powerful antioxidants which mop up many harmful free radicals are one of the most effective nutrients available for minimizing diabetic damage. **OPCs** protect against this damage by vastly improving the strength of the blood vessels and particularly by preventing tiny capillaries from rupturing. They are effective against a wide-ranging set of problems related to diabetes, including:

- Retinal damage (or diabetic retinopathy) due to tiny ruptured capillaries.
- Poor circulation, especially to the feet and lower limbs.
- Kidney damage or nephropathy.
- Excessive excretion of albumin.

As we have said before, OPC products come in the form of capsules and tablets and a daily dose of 50 to 100 milligrams is sufficient for the supplement to work effectively.

Alpha Lipoic Acid. This is not a vitamin but a coenzyme. Besides being a powerful antioxidant on its own, its presence enhances the effect of other antioxidants like vitamins C and E. Also known as **ALA** and **thioctic acid**, this coenzyme is especially helpful in reversing and preventing nerve damage due to diabetes. Due to its antioxidant power ALA cleans out harmful free radicals by itself, but the body converts some of the ALA into an even more powerful antioxidant known as **dihydrolipoic acid**. Together, both acids go after free radicals responsible for the development of atherosclerosis and certain neurological disorders. Alpha lipoic acid also helps in clearing out sugar from the blood besides coming to the aid of a faulty insulin mechanism.

ALA comes in the form of capsules. It should be used on an ongoing basis to be effective. Although some studies recommend a dosage of 600 milligrams a day, a dose of 200 milligrams will also be effective. For best results, take ALA in divided doses before meals. It is safe and without side effects when taken in the recommended doses.

Herbs. If one suffers from diabetes, herbs such as **buchu leaves, dandelion root, goldenseal** and **uva ursi** will prove helpful. In addition, **huckleberry** helps to promote insulin production while a tea made from **ginseng** can lower the blood sugar level.

Heart and Coronary Artery Diseases - Defended Against?

"Heart disease", used as a loosely applied and all-purpose term, covers a host of debilitating heart conditions. We will limit ourselves here to those far more common sets of symptoms, impairments and life-threatening conditions which are a direct fallout of the lifestyles we lead. What we're concerned about includes:

- Angina or angina pectoris
- Arrhythmias
- Cardiomyopathy
- Coronary artery disease
- Heart attack or myocardial infarction
- Heart failure and congestive heart failure
- Ischemic heart disease

All these conditions, their symptoms and the risk factors involved have been covered in depth in Chapter 10, so we will not repeat ourselves here. Suffice it to say that one can have heart disease and coronary artery disease without knowing it and the long process of deterioration may not be that apparent until a lot of damage has been done. Both maladies are serious conditions which only get worse if left untreated, ultimately resulting in an impaired quality of life or maybe even premature death.

TREATMENTS AND REMEDIES. The key to overcoming these debilities is to first know whether you are afflicted or not. Don't second-guess or make assumptions just because the symptoms are not overt where you're concerned! Get yourself checked out thoroughly because the good news is that these very serious conditions are amazingly responsive to management and intervention and there's every possibility of *successfully preventing* ***something*** *before it actually happens*!

Coenzyme Q-10. Also referred to as **cQ-10, ubiquinol-10** and **ubiquinone,** it is not really known whether this powerful antioxidant is actually a vitamin or simply operates as one. However, what everyone is pretty certain about is the fact that **coenzyme cQ-10**, an amazing substance found in small quantities in some foods and produced within the body, has a dramatic effect on heart health. The two important roles played by cQ-10 are, first, as an antioxidant it scavenges the free radicals that harm healthy cells and, second, as it kindles the chain reaction that allows cells to produce energy. The second, energizing role is especially important for the heart and, therefore, it's not surprising that healthy heart muscles normally have high concentrations of cQ-10. In fact there is a direct link between the lack of cQ-10 and heart disease with advanced heart failure cases evincing significantly less cQ-10 in heart tissue than cases exhibiting milder forms of the disease. Coenzyme Q-10 can dramatically reduce the symptoms of ***heart disease, congestive heart failure, angina or arrhythmia thereby improving the quality of life and even forestalling premature death***. Orally taking this supplement, which is well absorbed by the body in synthesized form, for as short a period as two months, will:

- Improve blood pressure, both systolic and diastolic.
- Improve lung function and reduce fluid accumulation in the lungs.
- Normalize heart rate.
- Reduce cyanosis, the bluish skin discoloration that signals lack of oxygen in tissues.
- Reduce fluid buildup (edema) in the ankles.
- Reduce shortness of breath, when at rest as well as during exertion.

Coenzyme Q-10 is generally available in capsule form. There are no clear guidelines regarding the exact dosage, but effective results were forthcoming with doses of between 50 and 150 milligrams per day. CQ-10 isn't a one-time cure and it may take a few weeks before any improvements begin to be felt. To obtain the full benefit, it must be taken on a daily, ongoing basis. The amino acid **carnitine** seems to enhance cQ-10's effectiveness. The supplement is safe and non-toxic and just 1.5 percent of cases have reported the side effects of nausea, upset stomach and/or diarrhea.

Vitamin C and Flavonoids. Vitamin C, as a powerful antioxidant, scavenges free radicals besides playing an amazing role in heart health by cutting the risks of coronary artery disease. Heart attacks are normally triggered by diseased arteries becoming blocked by accumulated plaque or clotted blood, being unable to accommodate the flow of blood and thus starving the heart of oxygen and nourishment. Vitamin C not only helps to control cholesterol and keeps blood platelets from clumping together in artery-clogging clots; it also strengthens and nourishes the tissues of the coronary arteries and the heart itself. Vitamin C also keeps the coronary arteries relaxed and able to dilate. **Flavonoids** are a large group of vitamin-like substances found in plants which, besides working independently to halt coronary artery disease, facilitate the body's use of vitamin C and boost its effectiveness in combating heart disease. Grape skin products rich in ***quercetin*** have been shown to promote relaxation and dilation of aortic tissues. The flavonoid ***rutin*** strengthens blood vessels. Combining ***rutin*** with vitamin C and ***hesperidin*** significantly increases blood flow in people with heart disease.

Both vitamin C and flavonoids are available in tablet form. Studies have shown that higher than average doses of vitamin C are required to combat heart disease and a daily dose of 2000 milligrams has successfully dilated clogged arteries. It is best to consult your doctor regarding the required dosage for your individual problem. It is also advisable to start off with smaller doses of vitamin C, gradually building up to the maximum daily dosage permissible. There is no recommended dosage for flavonoids and the best bet is to follow the package directions of the particular product purchased.

Carnitine. This amino acid plays a large role in heart health. Any deficiency in **carnitine** can damage the heart muscle tissue, while an adequate supply helps protect the heart from damage caused by decreased oxygen. Congestive heart failure interferes with the production of carnitine, making an already bad situation worse. However, taking the amino acid in supplement form has shown to help tremendously. The recommended dose is 900 milligrams a day and it is more effective when taken with cQ-10.

Herbs. For heart diseases the following herbs will help: **black radish, celandine, cheonanthus, fennel, ginkgo biloba, hops, horsetail, Irish moss, lobelia, parsley, red clover** and **rose hips**.

For coronary artery disease use: **blackberry, black cohosh root, cayenne pepper, ginseng, red grape vine leaves, Solaray #8 Formula** and **valerian root**. Patients benefit from **suma** herb tea of which three cups should be taken daily with **ginkgo biloba** extract.

Herbs common for both problems are **dandelion, butcher's broom** and **hawthorn,** which is the most effective herb of all. **Hawthorn** has strong antioxidant activities as it is rich in ***flavonoids*** and ***OPCs***. It has proved so effective as a heart protector that it has been recommended as "the first herb, besides garlic, that should be added to the daily dietary regimen". Hawthorn has been shown to:

- Reduce symptoms of congestive heart failure.
- Reduce angina
- Reduce arrhythmia and help to regulate heartbeat.
- Improve overall cardiac function and efficiency.
- Help keep coronary arteries relaxed and properly dilated, allowing more oxygenated blood to fuel the heart.
- Protect against future coronary disease.
- Improve stamina.
- Halt or retard future deterioration.

Hawthorn is available as capsules and drops. There is no recommended dose. If it is being used as a preventive measure, follow the manufacturer's label instructions. If one is being treated for heart disease, consult with a doctor.

High Homocysteine – Lower Away and Get a Stay!

As outlined in Chapter 2, **homocysteine** is a by-product of protein metabolism. This little-known substance is a risk factor for life-threatening heart attack and stroke. Unlike cholesterol, which may take years before it causes any serious problems, homocysteine works much faster and people born with the genetic inability to clear it from the blood, if they are not treated immediately, will develop severe atherosclerosis and die before the age of 15! When the body is functioning efficiently, homocysteine is usually converted into a harmless substance and quickly cleared from the blood. But there are many factors that can interfere with this process and cause the levels of homocysteine to increase which, in turn, can have a lethal effect on the body. It hardens and clogs arteries—including the arteries that feed the heart and the brain—and encourages the formation of blood clots in both veins and arteries thereby dramatically raising the chances of heart attack and stroke. Abnormally high levels of homocysteine should be a rare occurrence, but they aren't. The problem strikes around 33 percent of middle-aged and older people. ***Unfortunately, these high levels occur without any warning or overt signals and, to compound matters, neither do these high levels fall by themselves.*** The only way of knowing that one has high homocysteine is through an analysis of the blood. The following are some of the known factors that coerce homocysteine into running amok:

- **Deficiencies of Folic Acid and Other B Complex Vitamins**. It has been discovered that high levels of homocysteine are attributable to low intakes of **folic acid** (a B vitamin), **vitamin B6** and **vitamin B12**.
- **Genetic Inheritance**. Persons having a common genetic inheritance are put at high risk.
- **Alcoholism**. Active alcoholics have much higher levels of homocysteine than non-drinkers. A study revealed that when alcohol intake was totally stopped, the high levels dropped to reasonable readings within one or two weeks.
- **Coffee and Cigarettes**. Those habituated to nine or more cups of coffee a day are sure to have high homocysteine levels, as will those who smoke over 20 cigarettes daily. When both are combined—the less said the better!
- **Age**. As one grows older, the levels of homocysteine also rise automatically. It is yet to be ascertained whether this phenomenon is due to declining nutrition or the body's greater demands for certain vitamins or for some other reason.

What ***is*** very certain, is the fact that high levels of homocysteine take a correspondingly high toll on the circulatory system which can and will lead to a stroke, or a heart attack, or coronary artery disease, or blood clots in the veins and many other deleterious health problems. Fortunately, there is now a means of both halting and reversing this particular degenerative process—through the simple procedure of taking B vitamin supplements!

Folic Acid, B6 and B12. All three of these nutrients are members of the B vitamin family and are available in food sources, albeit most of us do not ingest enough for our own good! ***All three are considered essential for good health.*** Although B6 and B12 are no less important, it appears that folic acid plays the major role in controlling high levels of homocysteine. Since we never seem to be able to get enough of these vitamins through our food, we can now build up the needed quantities by taking them as supplements. The body utilizes supplemental B-complex vitamins in the same way that it utilizes the dietary forms or at least effectively enough to keep homocysteine levels from rising to abnormal levels. Medical science is confident that lowering the homocysteine levels with supplements of folic acid and vitamins B6 and B12 will prevent the onset of many problems that eventually lead to heart attack, stroke and premature death.

These B vitamins come in a wide array of tablets and capsules. They are available as individual vitamins as well as in combination (B complex) preparations. Since these work together and compliment each other, it is important to take all three. The most convenient method of doing this is by taking the B complex capsules. What should be ensured is that a minimum of between 500 to 750 micrograms of folic acid is being consumed daily through these combinations. High doses of these vitamins are likely to cause gastric upset, gas, bloating and diarrhea.

Hypertension and Stress - The Silent Killers!

High blood pressure has been vividly described in Chapter 3—the causes—and Chapter 10—the symptoms and diagnostic uses. We are not going down that descriptive route again, but we do request that you flip back for a refresher peek. It is a documented fact that sixty million Americans suffer from high blood pressure or hypertension, in one form or another, and it is has become the biggest medical challenge of this century. Unfortunately, because of the extreme diversity of India's population, it is not possible to ascertain the exact figures for our country, but you can bet your bottom dollar that the actual number will outmatch the American figure by at least another 100 percent! If you, along with many of your fellow Indians, happen to belong to this band of "hyper-tensioners", your risk of heart attack is three times greater than and your risk of stroke double that of persons with normal pressure.

Basically, the malady presents itself in two forms: primary, or essential, hypertension and secondary hypertension. The latter form accounts for only about 15 percent of all cases and is normally brought about by an identifiable cause such as kidney disease or a tumor. Once the cause has been treated and eradicated, or brought under control, the blood pressure does return to normal. What really bothers the medical fraternity is primary hypertension, which presents itself without any specific cause and for unknown reasons. In such cases, to bring the pressure within normal limits, there really is no alternative but to resort to drugs, and the higher the pressure the higher will be the drug dosage. This usually results in a host of side effects, such as fatigue, depression and possible impotence, with which the patient must also contend. In addition, there may be some elderly patients who are on other types of medication besides blood pressure drugs and this leaves them open to the risk of drug interaction.

The force exerted by blood flow against the blood vessel walls is referred to as blood pressure and the reading contains two numbers. ***Systolic pressure,*** or the pressure when the heart contracts to empty its blood into the circulation, is denoted by the higher number. ***Diastolic pressure***, or the pressure when the heart relaxes to fill with blood returned by the circulation, is the lower number. The results, in millimeters (mm) of mercury, are reported as a ratio expression of maximum over minimum, for example, 140/80. In healthy adults, the value at age 30 is around 120/80 and increases to about 140/85 at age 40 and over. This increase occurs as the arteries lose the elasticity that, in younger people, absorbs the force of the heart contractions. However, blood pressure is influenced by many factors and varies among individuals as well as for an individual at different times. If one already has high blood pressure and is on medication, taking these should be regular as should the visits to your doctor.

TREATMENTS AND REMEDIES – DROP THE PRESSURE! It is recommended that the following measures be adopted and after a sufficient trial period—say a month or two—get a pressure reading taken which will indicate whether the newly initiated steps are having the required effect. Often, your physician may suggest that you purchase a sphygmomanometer— commonly called a blood pressure cuff—to measure your blood pressure at home on a regular basis (weekly is typical) to monitor your "drop the pressure" progress.

Diet. If you are overweight, weight loss is the most effective means of reducing your blood pressure and this will only come about if your diet is changed. Change of diet doesn't mean going off food! It's more a matter of regulating your intake and eating only those foods that are good for overall health. What this actually means—get rid of junk food and avoid saturated fats like the plague!

Vegetarian Diet. The components of the vegetarian diet, i.e., more fibre, vitamin C, vitamin E, magnesium, calcium and potassium and significantly less salt, total fat, saturated fat and cholesterol, definitely help to fight hypertension. Most vegetarians enjoy lower blood pressure readings than those who are basically non-vegetarians. We're not advocating here that all patients necessarily become vegetarians because of this reason. In fact, there are many who do not do well on a vegetarian diet and may feel tired and run-down. It is also impractical to become a strict vegetarian. While strict vegetarianism may not be sensible, patients are advised to adopt as many aspects of a vegetarian diet as possible, particularly the reduction of saturated fats and cholesterol.

Sodium and Salt. Today we know that not everyone with hypertension is salt sensitive and that old adage of eliminating salt from your diet, because of high blood pressure, is not strictly true. In fact, only 30 to 40 percent of the population is salt sensitive. If you have a heart or kidney problem in addition to high blood pressure, you should definitely avoid salt because your body does not eliminate sodium properly. If you are restricting sodium solely for hypertension, you might want to test yourself to see if sodium restriction is doing you any good. After having your pressure taken, follow a diet that is sodium free for two weeks and then have your pressure taken again. You can do the opposite if you have been avoiding salt i.e., try two weeks of a diet that includes salt and see what effect, if any, it has on your pressure readings. If you are salt sensitive, you should probably restrict your salt intake and increase your potassium intake. Many people who are salt sensitive not only consume too much salt but also get too little potassium in their diet. This causes an increase in fluid volume and an impairment of the blood pressure regulating mechanism. You can either rely on potassium supplements or you can make a concerted effort to get it from food sources. Many patients were able to reduce their medications by consuming a diet rich in potassium. Good sources are fresh fruits and vegetables such as potatoes, peas, peppers, eggplant, pears, lima beans, tomatoes and bananas.

High Fibre Diet. Just as a high fibre diet has the amazing effect of lowering high blood pressure, diets that are low in fibre definitely contribute to the onset of hypertension. This does not necessarily mean that one should resort to taking fibre supplements. The benefits accrued by resorting to diets of fresh vegetables, fruits, whole grain cereals, breads and pastas cannot be overemphasized. It has been noted that a bowl of oat bran cereal or other high-fibre cereal eaten at breakfast has definitely helped and is highly recommended.

Celery. It lowers blood pressure by relaxing the smooth muscles in the blood vessels themselves. Although celery is often forbidden to people with hypertension because of high sodium content, its beneficial aspects have now been recognised. It is worth trying out an intake of four celery stalks daily for a three to four week period. This will provide sufficient amounts of the active chemical which causes the desired effect. Do not overdose because celery can be toxic in large amounts.

Alcohol. If you consume alcohol, you are more than likely to get hypertension. Despite the reports that *moderate* alcohol intake has a certain protective effect, patients with any degree of hypertension are advised to completely eliminate alcohol from their diet.

Calcium and Magnesium. These two minerals work together in the body and low levels of both are associated with high blood pressure. Calcium's interaction with magnesium, and particularly with sodium, may affect blood pressure and studies have revealed that some patients have benefited by taking calcium and magnesium supplements. It is suggested that one may take both calcium and magnesium supplements on a trial basis for two months and, if a reduction in blood pressure is seen, these may be continued. If no significant drop in pressure occurs, discontinue taking the supplements. Recommended daily doses are 1000 milligrams of calcium at bedtime and 250 milligrams of magnesium twice a day for a trial period of two months.

Coenzyme Q10. As already described, this supplement has been extremely helpful for many people with cardiovascular disease including hypertension. It improves the function of the blood vessel wall and thus helps regulate blood pressure. Take 30 milligrams three times daily.

Garlic. It has been shown to decrease the systolic pressure by 20 to 30 mm Hg and diastolic pressure by 10 to 20 mm Hg. Patients are advised to take two to three cloves of garlic daily to obtain therapeutic benefits and enjoy garlic in food as often as possible. You can even take odorless garlic which is available in capsules.

Vitamin C. This vitamin, so useful for many metabolic functions, has been found to help regulate blood pressure. Take 1000 milligrams daily.

Exercise. This can be effective in treating hypertension. It not only has physiological effects but also reduces stress, therefore offering enormous benefits to the cardiovascular system. Patients suffering from hypertension are advised to take a brisk half-hour walk three to four times a week.

A Healthy Liver – The Secret of a 'Long-Liver'!

We met up with this 'big boss' of our internal detoxification plant in Chapter 2. The sad part is that the importance of the liver generally tends to be overshadowed by the importance given to the heart. As we have already mentioned, every organ in our body, besides being inexorably linked functionally, has equal importance in the sustenance and prolongation of life. The liver, just as essential for bodily functions and as vulnerable to "attack" as any other organ, performs more than a dozen important operations. These functions can be briefly summarized as:

- Boosting the immune system to aid resistance to diseases
- Cleansing the blood of bacteria and waste products
- Helping to regulate blood clotting
- Maintaining hormone balance
- Manufacturing proteins
- Metabolizing alcohol
- Monitoring and regulating the blood levels of many chemicals and drugs
- Neutralization and destruction of toxins
- Producing quick energy on demand
- Production of bile which is essential for proper digestion
- Regulating fat transportation
- Regulating the production and excretion of cholesterol
- Replacing its own damaged tissue
- Storage of certain vitamins, minerals and sugars
- Storage of iron

The liver, therefore, is definitely the body's main waste-filtering organ, targeting and removing bacteria and other organic toxins from the blood.

Nowadays, the liver has the additional job of also dealing with innumerable inorganic wastes that somehow find their way into the body. We're talking here about chemicals from air pollution, cosmetics, aerosols, synthetic fibres, cleaning fluids, dyes and a whole host of other modern-day products. Even a few of the beneficial medications are sometimes capable of harming the liver. So, how does one figure out whether his or her liver is healthy? Though many do not have liver disease, their livers are still subject to slow and significant damage over a period of time—even one's daily routine puts the liver under all sorts of stress! The liver can be seriously threatened by a number of diseases, the common ones being:

(a) Fatty Liver. This is a condition where fat accumulates in the liver. Though it is not a disease in the true sense and does not of itself impair the liver's function, a fatty liver is a symptom of other health problems and an indication of worse troubles in store. The prime causes are alcohol consumption, extreme overweight and diabetes. When these problems are properly treated, the condition normally gets better on its own.

(b) Hepatitis. There are several distinctly different forms of this inflammatory liver disease. Of the various types of viral hepatitis, each is caused by a different virus, each has a different transmission route and the outcome of each is also different. Some prominent forms are:

- ***Alcoholic hepatitis.*** This form is caused by heavy drinking
- ***Chronic hepatitis.*** This form is caused by viral infections, medications, metabolic disorders or immune system disorders.
- ***Hepatitis A.*** This is spread through fecal matter, through contaminated shellfish, through anal sex and through kissing. This is easy to cure and has no lasting effects.
- ***Hepatitis B.*** This condition is far more serious and is spread through various bodily fluids such as saliva, blood, semen, vaginal secretions, tears and breast milk. Though most patients do stage a recovery, a small percentage becomes permanent carriers of the virus. An even smaller percentage will eventually develop serious liver problems such as chronic liver disease, cirrhosis and cancer.
- ***Hepatitis C.*** This form is transmitted through blood. Since stringent norms are in force regarding the testing of donated blood, the risks connected with transfusions have mostly been

eliminated. People at risk these days are drug users who share needles and those who indulge in unprotected sex.

(c) Cirrhosis. This chronic condition exists when normal liver cells are damaged and the healthy cells are replaced by scar tissue, thus making it impossible for the liver to function efficiently. Though over-consumption of alcohol is the main cause, cirrhosis can also be brought about as a result of chronic hepatitis, diabetes, inherited diseases and conditions, bile-duct diseases, long-time exposure to a toxic environment, a severe adverse drug reaction or parasitic infections.

(d) Cancer. This disease does not commonly originate in the liver. However, contracting hepatitis B or C increases the risk of developing cancer of the liver.

Although there may be varying symptoms depending on which type of liver disease has been contracted, the general debilitating signs are:

- Dark-coloured urine.
- Drastic changes in the colour of the stool.
- Fatigue, loss of stamina or loss of sex drive.
- Generalized itching that does not go away.
- Jaundice, evident by the skin or the whites of the eyes turning yellow.
- Loss of appetite, nausea and/or vomiting.
- Passing or vomiting blood.
- Swelling of or pain in the abdomen.
- Unexpected weight loss or gain.

TREATMENTS AND REMEDIES – SO, WHAT'S TO BE DONE? Most problems can be prevented and overcome by correctly identifying the root cause and treating the malady accordingly before too much damage has been done. A lot of those who may be well informed about cardiac and some other diseases are woefully lacking in knowledge about the liver and tend to take this overworked organ for granted. Similar to the heart, the liver has an extremely stressful job and needs constant attention and consideration. If looked after properly, the liver is capable of giving that vital life extension when the need so arises.

Milk Thistle. This is considered a weed quite widely; however, the plant contains **silymarin**, a flavonoid group that has a natural affinity for the liver and works in multiple ways to protect and heal. Though the young leaves are soft and edible, they lack a concentration of the required nutritional substances. Therefore, extracts, in capsule form, made from the fruit of this plant are more potent. The benefits of milk thistle extract on the liver include the following:

- **Protects against toxins and pollutants**. Though the method of this occurrence is still not fully understood, milk thistle's active substances make liver cell walls unresponsive to toxic machinations.

- **Stimulates regeneration of liver tissue**. Once damaged, liver tissue takes an inordinately long time to replace itself. Milk thistle improves the liver's functioning ability by enhancing the tissue regeneration rate.

- **Offers a better outlook for hepatitis patients**. Milk thistle is very effective in treating both hepatitis B and chronic hepatitis. The levels of the particular enzymes which indicate liver damage are decreased and overall liver function improves.

- **Improves cirrhosis conditions**. Milk thistle has dramatically improved the overall condition of cirrhosis patients. Its amazing ability to improve functioning in a damaged liver has even allowed patients with diabetes-related cirrhosis to reduce insulin intake.

- **Protects against cancer**. Among its other constituents, silymarin also contains an amazing and very potent antioxidant called **silybin** which consumes cancer-promoting free radicals.

- **Increases glutathione levels**. Another powerful antioxidant and natural cancer fighter is **glutathione**. Silymarin increases the levels of glutathione in the liver.

Concentrated milk thistle is available as a liquid extract as well as in the form of 200 milligram capsules standardized to contain 70 to 80 percent ***silymarin***. An effective average dose is three such capsules daily. The side effects (limited to gastrointestinal upsets such as nausea or an upset stomach) are mild, rare and temporary.

Herbs. All types of hepatitis can be treated effectively with **black radish, dandelion, goldenseal** and **red clover.**

Ayurvedic Preparation. A locally manufactured ***ayurvedic preparation***, called **Livosin**, is readily available in liquid form and quite effective in the treatment of hepatitis. The decoction is made up and contains herbs such as **kalmegh, hansapadi, bhringaraj, ajowan, hing, pudina** (mint), **adrak** (ginger), **sonapata, triphala, carica papaya, anantamool, aswagandha, jau, arjun, biraga, kurchi, bhuiamlaki, dhania, pitt papda, gulancha, ash-shawra, yasthimadhu** and **ghritakumari.** An effective dosage is 2 teaspoons twice daily.

Osteoporosis – No Longer the 'Bone Fragilizer'!

As ageing progresses and with the onset of menopause, falling estrogen levels as well as the body's hampered ability to efficiently absorb calcium leaves the system in a state of chronic calcium deficiency. This is a major problem as calcium is an essential component of several body fluids and, to make good the "loss", the body reacts by pulling calcium from its own bones. This action of depleting calcium in the bone structure results in the bones becoming brittle, fragile and easily broken. Loss of estrogen leads to rapid bone loss which creates the disease known as **osteoporosis.** This is not a disease that anyone can catch as it's a condition with well-established risks. The chances of bone loss are influenced by the following factors:

- **Being a Woman**. Bone loss is far more rapid and dramatic in women than in men, especially during menopause, or premature menopause brought on by a hysterectomy or some other procedure.
- **Age**. Osteoporosis is age-related in both men and women.
- **Family History**. It is medically proven that osteoporosis has a genetic link.
- **Diet.** Most current-day diets contain insufficient calcium.
- **Bone Structure**. Persons with heavier and denser bones are less likely to suffer from osteoporosis than are those with delicate frames.
- **Taking Steroid Drugs**. Use of steroid drugs over prolonged periods can make one susceptible to osteoporosis. Medication for rheumatoid arthritis, inflammatory bowel syndrome and some other diseases also poses the risk of osteoporosis.

Other causes of this disease are an inability to absorb sufficient amounts of calcium through the intestines, a calcium-phosphorus imbalance, prolonged jaundice, a gastrectomy, a lactose intolerance and lack of exercise. The most visible symptom of osteoporosis is a declining or stooped stature. This happens because the bones of the spinal column weaken, cannot support the upper torso and a stoop or curvature of the spine results. There are many worse happenings inside the body as the bones get progressively porous and brittle. Knocks and falls that would have been of little consequence in younger years increasingly present the risk of fractures.

SO, HOW DO I PREVENT ME FROM "BREAKING IN HALF"? Osteoporosis is progressive unless some action is taken. The condition was once thought to be an inevitable part of ageing, but not any longer! By initiating some simple preventive steps, one can ensure that he or she will still stand tall in their golden years.

Calcium. This is the common mineral used for building bones and teeth, but there are many people who don't get enough calcium even in the best of circumstances. This mineral also plays a role in transmitting nerve messages and regulating heartbeat. As one ages, the need for calcium also increases as the body does not absorb it as efficiently as before. In the context of calcium absorption through natural means, modern diets are almost as totally lacking in calcium as is the general lack of knowledge regarding the importance of this vital and indispensable nutrient that is necessary throughout the life cycle. Since we do not normally take in enough calcium with our food, we must necessarily find some other means of obtaining it. Medical research has revealed that calcium in supplement form is as effective for osteoporosis as is added dietary calcium. Supplementation has shown to:

- Preserve bone density
- Dramatically slow bone loss throughout the body
- Totally eliminate the loss in some bones
- Bring about a reduction in the number of fractures.

Calcium is available in tablet form and is often combined with magnesium and zinc, which helps the body to better use calcium. An intake of 1000 to 1500 milligrams of calcium a day will be sufficient to prevent osteoporosis from setting in. The body makes best use of calcium when food is present in the digestive system, so the supplement should be taken with meals. Calcium is completely safe and non-toxic

Vitamin D. This vitamin is necessary for proper calcium absorption and the lack of vitamin D has been linked to low calcium levels. The best source of vitamin D is sunlight; however, this also poses the risk of skin cancer. The best method of ensuring the required intake of sufficient vitamin D (400 IUs must be taken daily) is to resort to fortified foods and/or a vitamin supplement.

Herbs. Feverfew, horsetail (silicon), **oatstraw** (a form of silica) and **shavegrass** have all proved effective in cases of osteoporosis.

Enlarged Prostate – An Unnecessary Nuisance!

The prostate gland, part of the male reproductive system, sits between the rectum and the bladder and surrounds the proximal portion of the urethra. Even after a man reaches physical maturity, the prostate continues to grow so gradually that the growth goes unnoticed for decades. However, sometime around middle age, problems often arise when this enlarged walnut-shaped gland presses in on the urethra which it surrounds. This condition is called **Benign Prostate Hypertrophy (BPH).** Generally, the condition of an enlarged prostate occurs between the ages of 50 and 60 years, and studies indicate that the percentage of men afflicted with an enlarged prostate is categorized as follows:

- 80% above the age of 80 years
- 50% between the ages of 70 and 80 years
- 25% between the ages of 60 and 70 years

Symptoms. The urine is neither passed in a strong flow nor is there the requisite force to eject it fully and sometimes there may be a considerable waiting period before the urine flow actually starts. As a result, some residual urine may be retained which could result in retrograde pressure in the kidneys further causing partial impairment of renal function, a rise in the levels of urea and creatinine and the possibility of cancer taking hold. Other symptoms include (1) the sudden urgency to urinate and (2) nocturia—the need for frequent urination during periods of sleep. Previously, surgery was considered the only and ultimate remedy, but lately certain discoveries of advanced homeopathy have been successfully used in treating persons having hypertrophy of the prostate.

There are two types of benign diseases of the prostate, namely:

1. **Non bacterial type.** This type is the most common and the complaints are usually (1) lack of force in passing urine and (2) perineal or suprapubic discomfort, and there may also be irritative urine.

2. **Bacterial type.** This type occurs when the prostate swells up and obstructs urinary flow. Fever and irritation while passing urine may be common complaints. The urine culture will show bacteria, pathogens and pus cells.

It has been observed that practitioners/physicians generally advise most of their patients to drink lots of water or other liquids. Contrary to this, ***research has revealed that excess intake of fluids/ liquids will cause distention in the bladder which holds back the urine because the outlet passage is being obstructed by the hyperplasia of the prostate.*** This causes pressure in the kidneys leading to renal failure and bacterial infections. Depending upon other clinical findings, patients should be advised to drink water only to quench thirst.

Prostate Cancer. This cancer is discussed here rather than in the CANCER – STILL DEADLY BUT NEGOTIABLE section, preceding, because it has ***symptoms similar to BPH.*** Therefore, investigations ***must*** be carried out to assess if one has cancer. Detection techniques such as the transrectal ultrasound test and the **prostate specific antigen (PSA)** blood test should be undertaken. This screening effort should be directed to detection and treatment of only those prostatic carcinomas most likely to cause morbidity and mortality if left untreated. As a first line screening test, PSA will be elevated in approximately 10-15% of men with self referred screening. Approximately 20-25% of men with intermediate degrees of elevation (4.1-10 ng/ml (nanogram per millilitre) with normal being less than 4.0 ng/ml) will be found to have prostatic cancer. When the PSA is in the intermediate degree of elevation, biopsy samples will be taken. Almost two-thirds of those with elevations greater than 10 ng/ml will have cancer. Patients with intermediate levels of PSA will usually have localized and therefore potentially curable cancers.

TREATMENT AND REMEDIES. There are several treatments that can help a benign enlargement of the prostate gland. Therapies for treatment of prostate cancer are not discussed here.

Non-Bacterial Prostatitis. Most cases are helped by **sarsaparilla 30** taken once a day. The relief initially fluctuates between complete and partial. However it has to be continued for several weeks to a few months for complete recovery. In some cases there may be a burning sensation while passing urine. Then patients may be given **medorrhinum 200** at intervals of three days when sarsaparilla temporarily ceases to act.

Bacterial Type. Acute stage of febrile condition with pain in the sacral and perineum area, pus cells and bacteria in urine is helped by **thuja 30** taken every three hours which controls the pus and fever. When there is distress and obstruction in passing urine, sarsaparilla 30 given alternatively with thuja 30 gradually improves the flow of urine.

Saw Palmetto. This herbal member of the palm family is available in tablets and capsules. It is believed that men with an enlarged prostate have elevated blood levels of ***dihydrotestosterone*** (DHT), which is a powerful sex hormone. Saw palmetto lowers DHT and also has an anti-inflammatory effect. Take 320 milligrams as a daily dose on a trial basis for a four to six week period. If it gives relief, then continue.

Pygeum Africanum. This herb is derived from the bark of the pygeum africanum tree that grows in the mountains of southern Africa. It is a powerful anti-inflammatory that functions to reduce swollen prostrate tissue. The most outstanding characteristic of pygeum is that when combined with saw palmetto, the combination is much more effective than either herb taken individually. The minimum recommended dose is 100 milligrams per day (standardized to contain 13 % sterols) with some practitioners recommending up to 200 mg/day. Be sure and check the pygeum content if using a commercially prepared "prostrate herbal formula" since for many brands the pygeum content is *much* less than the 100 mg recommended daily dosage.

Herbal Teas. Acute inflammation or enlargement of the prostate gland will often respond to certain herbal teas:

- Make a decoction of equal quantities of **gravel root, sea holly** and **hydrangea root**. Three to four tablespoons of this taken three times daily will ease inflammation and reduce the discomfort of urination. **Marshmallow leaves,** for their demulcent properties, may be added to this mixture if burning and heat persist.

- Other helpful teas can be made from the diuretic herb **buchu** (do not boil) and from **corn silk**.

Horsetail is an astringent and can be used in cases of small amounts of blood being passed as well as to treat frequent urination at night. Combine it with **hydrangea** for greater effect on an enlarged gland. Other helpful herbs for these symptoms are **goldenseal root, parsley, juniper berries, uva ursi, slippery elm bark** and **ginseng**.

Enhanced Health Within Easy Reach?

You bet! But just as long as blinders are not restricting our vision! Throughout this account you will have noticed that the accent is on cures but the stress is on good health. It is quite obvious that being in good health automatically ensures that the body is more likely to be able to fend off disease and malaise. Unfortunately, there is nothing automatic involved about ***maintaining*** good health, which not only requires our unrequited dedication for its preservation but also our total subjugation to the laws of nature. That old military adage that "the more we sweat in peace, the less we bleed in war" is certainly apt in this continued conflict between health and disease—a conflict in which disease appears to have the upper hand and superior weapons. In addition to the alternative therapies, medicines and treatments we have highlighted, our only real weapon against disease is *knowledge*. Unless we know all about the various maladies and the debilitating effects they have on the human body ***AND*** take concrete steps to overcome adversity, we will always be on the receiving end of whatever disease decides to dish out. As we have cautioned, the number of new ailments discovered almost daily is far higher than the number of cures discovered for the maladies that we're already living with! Far be it for us to comment on what mankind has in store for itself on the health front, but what is truly amazing is that the ways and means to good health and good living are but a handclasp away—something that mankind excels in failing to grasp! Your life is literally in ***YOUR*** hands, but do not reject or turn down the helping hand that alternative therapies unstintingly offer in this respect. Let us do our bit by first ensuring our personal good health. It is only then that we can pass on "good health" to others and to our planet.

INDEX

– A –

— B —

— D —

— E —

– F –

– G –

– H –

– R –

– S –

– T –

– U –

– V –

– X, Y & Z –

Printed in the United States
86733LV00002B/112-135/A

9 780971 991606